7th International Conference on Actual Problems in Machine Building 2020

AA007547

IOP Conference Series: Materials Science and Engineering
Volume 843

Novosibirsk, Russia
25 March 2020

ISBN: 978-1-7138-2684-2
ISSN: 1757-8981

Printed from e-media with permission by:

Curran Associates, Inc.
57 Morehouse Lane
Red Hook, NY 12571

CURRAN ASSOCIATES INC.
proceedings
.com

Some format issues inherent in the e-media version may also appear in this print version.

Institute of Physics
Dirac House, Temple Back
Bristol BS1 6BE UK

Phone: 44 1 17 929 7481
Fax: 44 1 17 920 0979

techtracking@iop.org

Additional copies of this publication are available from:

Curran Associates, Inc.
57 Morehouse Lane
Red Hook, NY 12571 USA
Phone: 845-758-0400
Fax: 845-758-2633
Email: curran@proceedings.com
Web: www.proceedings.com

TABLE OF CONTENTS

IOP Conf. Series: Materials Science and Engineering **843** (2020) 011001 doi:10.1088/1757-899X/843/1/011001

PREFACE

The 7th International Conference on Actual Problems in Machine Building was held at Novosibirsk State Technical University on 25 March.

Originally the conference was supposed to be an academic discussion within the course of a major industrial exposition MashExpo-Siberia. The exposition is dedicated to novel tools, equipment and technologies for metalworks and mechanical engineering. Unfortunately, the exposition was postponed to an autumn due to the Covid pandemic. According to a current schedule MashExpo-Siberia will be held on 15-18 September 2020.

Despite the fact that the exposition was postponed, small scale scientific meeting took a place in the conference hall in Novosibirsk State Technical University. Not all the manuscripts included in a programme were reported at the session. However, some reports were virtually made via Skype. We believe that the more studies will be presented at MashExpo 2020 this September.

This volume contains 27 reported studies dedicated to such topics as

- innovative technologies in mechanical engineering,
- technological equipment, machining attachments and instruments,
- materials science in machine building.

The conference helped local Siberian mechanical engineers to get acquainted with the research and inventions that carried out in many regions of the country and abroad. All papers were reviewed by two referees. The main criteria for selecting articles were the relevance of the problem and the novelty of the proposed solution.

With the kind regards,
Chair of the Committee
Vadim Yu. Skiba

IOP Conf. Series: Materials Science and Engineering **843** (2020) 011002 doi:10.1088/1757-899X/843/1/011002

Peer review statement

All papers published in this volume of *IOP Conference Series: Materials Science and Engineering* have been peer reviewed through processes administered by the proceedings Editors. Reviews were conducted by expert referees to the professional and scientific standards expected of a proceedings journal published by IOP Publishing.

IOP Conf. Series: Materials Science and Engineering **843** (2020) 012001 doi:10.1088/1757-899X/843/1/012001

Design of surface profile of pairs of friction unit

P A Polyakov, A E Litvinov, E A Polyakova, E S Fedotov and R S Tagiev

Kuban State Technological University, Krasnodar, 2a, Moskovskaya st., Krasnodar, 350020, Russia

E-mail: artstyleone@mail.ru

Abstract. To control the processes of friction and wear in the working pairs of the friction unit, it is necessary to study the behavior of the materials of the friction unit during contact interaction, accompanied by the combined action of temperature and dynamic loads. Numerous studies are devoted to the study of the influence of the chemical composition on the friction process of friction units, as well as the creation of new materials for brake units. There are practically no works devoted to the influence of nanogeometry of the surfaces of contact pairs of friction units on contact pressure. In this case, from an uneven distribution of contact pressure over the friction surface, strong heating can occur, which results in the formation of burns, thermal spots and foci of microcracks.

1. Introduction

The researchers [2,3,4] indicate the effect of temperature on the tribocontact of various friction units. In literature [6,7], modified friction units based on new polymeric materials used in various types of brake devices are proposed. In the source [5], the author cites the effect of hydrogen on the interaction process of friction pairs. Literary sources [1,8] describe the interaction of electric fields of the surfaces of contact pads of friction units. The analysis of the works [9, 10] on the wear of the working surfaces of friction units shows that the specific contact pressures of the working surfaces of friction pairs are unevenly distributed over the area of the metal element.

From the literature review, we can conclude that the authors paid great attention to temperature, chemical composition as the primary factor that affects the contacting elements of the friction unit. The consequence of this phenomenon is uneven wear of both metallic and non-metallic friction elements. However, the temperature is the result of contact and partial smoothing of the microprotrusions of non-metallic and metallic elements. The influence of the distribution of nanogeometry over the contact surface of the friction unit is not considered in any source. On this basis, we can conclude that this problem is not sufficiently disclosed in modern science.

To solve the formulated problem, it is necessary to formulate a requirement for the designed friction units, which relate to the minimum uneven distribution of pressure over the area of the working surfaces of the brake. The durability of the friction unit parts depends on the uniformity and wear rate of the working surfaces. The main cause of uneven wear is the contact pressure of the friction pairs.

The problem of choosing nanogeometry of the surface of the working surfaces of the friction unit, which ensures uniform wear of the metal friction element, has not yet been solved by the calculation method.

Based on the choice of nanogeometry of the friction surface, it is possible to obtain a uniform law

IOP Conf. Series: Materials Science and Engineering **843** (2020) 012001 doi:10.1088/1757-899X/843/1/012001

of the distribution of contact pressure over the elementary area of friction of the friction assembly, which will further reduce wear on working surfaces.

The value of the optimal value of the contact pressure of the elementary area on the friction surface $\bar{p}$ is not known in advance and must be determined when solving the optimization problem.

In addition to wear, the uneven distribution of contact pressure over the contact area of the friction unit affects the occurrence of thermal foci on the surfaces of the metal elements of the friction unit (Figure 1). All these factors naturally do not contribute to the effective functioning of such a critical unit as the braking mechanism.

Figure 1. Thermal stains on the surface of the brake disc

2. Materials and methods

To solve the contact problem, it is necessary to create a mathematical model based on the differential equations of thermoelasticity with boundary conditions. The control variables are the data of nanogeometry of the friction surfaces, i.e. external surfaces of metallic and non-metallic elements of the friction unit. The elements of the friction unit are modeled as an isotropic homogeneous body. We place the elementary area of the friction surfaces in the polar coordinate system. The origin is located in the center of the concentric arcs L0 and L with radii R0 and R. We assume that the contour of the nonmetallic friction element L 'is close to circular in which the function H (θ) lies in the definition of a further solution to the optimization problem. Imagine the boundary of the contour of a non-metallic friction element in the form:

$$\rho = R + \varepsilon H(\theta), \tag{1}$$

where ε is a small parameter equal to the ratio of the maximum height of the profile roughness to the average value of the height of the friction surface.

Similarly, the external contour of the elementary area of the metal friction element is presented, in which the function H1 (θ) is determined when solving the optimization problem:

$$\rho = R_1 + \varepsilon H_1(\theta), \tag{2}$$

where R1 is the outer radius of the elementary area of friction of the metal friction element, m

Before solving the optimization problem, it is necessary to solve the contact problem of pressing a non-metallic element into the working surface of a metal element, considering the functions H (θ) and H1 (θ) known, and each function has the form of a segment of the Fourier series.

The contact task problem has the form of a system of equations:

IOP Conf. Series: Materials Science and Engineering **843** (2020) 012001 doi:10.1088/1757-899X/843/1/012001

$$\begin{cases} p^{(1)}(\theta,t) = p_0^1(\theta) + tp_1^1(\theta) + \ldots + tp_n^1(\theta); \\ p_0^{(1)}(\theta) = \alpha_{0,0}^1 + \sum_{n=1}^{\infty}\left(\alpha_{k,0}^1 \cos n\theta + \beta_{k,0}^1 \sin n\theta\right) \\ p_1^{(1)}(\theta) = \alpha_{0,1}^1 + \sum_{n=1}^{\infty}\left(\alpha_{k,1}^1 \cos n\theta + \beta_{k,1}^1 \sin n\theta\right) \\ \alpha_n^{(1)} = \alpha_k^1 + \alpha_{n,1}^1 t + \ldots; \ \beta_n^{(1)} = \beta_k^1 + \beta_{n,1}^1 t + \ldots; \end{cases}, \quad (3)$$

where t is the contact time of elementary sites, s.

According to the system of equations of the contact problem, the pressure of microroughness linearly depends on the coefficients of the Fourier series $\alpha_{k,0}^1$ and $\beta_{k,0}^1$of the function H (θ) and the coefficients $\alpha_{k,1}^1$ and $\beta_{k,1}^1$ of the function H1 (θ).

For the elementary contact pad of the friction unit, the contact pressure is represented as:

$$p(\theta,t) = f\left(\theta, t, \alpha_{k,0}^1, \beta_{k,0}^1, \alpha_{k,1}^1, \beta_{k,1}^1\right), (k = 1,2, \ldots m); \quad (4)$$

Contact time is considered a free parameter. Coefficients from the Fourier series $\alpha_{k,0}^1; \beta_{k,0}^1; \alpha_{k,1}^1$ and $\beta_{k,1}^1$ are constant and must be determined. First it is necessary to split the segment [θ1, θ2] of the variable θ into M parts, where M> 4m + 2

$$p(\theta_i,t) = f\left(\theta_i, t, \alpha_{k,0}^1, \beta_{k,0}^1, \alpha_{k,1}^1, \beta_{k,1}^1\right), (i = 1,2, \ldots M); \quad (5)$$

$$\theta_i = \theta_1 + \Delta\theta_i; \qquad \Delta\theta_i = \frac{\theta_2 - \theta_1}{M};$$

It is necessary to find the values of unknown parameters, which will optimize the function of contact pressure. Deviations of the contact pressure of the elementary friction pad from the desired constant p¯ are calculated by the formula:

$$\varepsilon_i = f\left(\theta_i, t, \alpha_{k,0}^1, \beta_{k,0}^1, \alpha_{k,1}^1, \beta_{k,1}^1\right) - \bar{p}, (i = 1,2, \ldots M); \quad (6)$$

The basic principle of the least squares method states that the values of the contact pressure of the elementary friction site of the friction unit will be most likely, provided that the sum of the squared deviations ε_i is minimal:

$$U = \sum_{i=1}^{M}\left[f\left(\theta_i, t, \alpha_{k,0}^1, \beta_{k,0}^1, \alpha_{k,1}^1, \beta_{k,1}^1\right) - \bar{p}\right]^2 \to \min; \quad (7)$$

At any time, the coefficients $\alpha_{k,0}^1, \beta_{k,0}^1, \alpha_{k,1}^1$ и $\beta_{k,1}^1$ (k=1,2...m) and $\alpha_{k,0}^1, \beta_{k,0}^1, \alpha_{k,1}^1$ и $\beta_{k,1}^1$ (k=1,2...m) and the quantity the optimal values of the contact pressure of the elementary area on the friction surface p¯ are considered as independent variables and, equating the partial derivatives of the left side with respect to these variables to zero, we obtain exactly 4m + 3 equations with 4m + 3 unknowns. Since the function $f\left(\theta_i, t, \alpha_{k,0}^1, \beta_{k,0}^1, \alpha_{k,1}^1, \beta_{k,1}^1\right)$ is linear with respect to unknown parameters, then the compilation and solution of the system is simplified.

The contact pressure at the elementary friction site of the friction unit is determined by the formula:

$$p(\theta,t) = p^{(0)}(\theta,t) + \varepsilon p^{(1)}(\theta,t); \quad (8)$$

The function $p^{(0)}(\theta,t)$ does not depend on the desired coefficients, and the function $p^{(1)}(\theta,t)$), on the contrary, linearly depends on the coefficients $\alpha_{k,0}^1, \beta_{k,0}^1, \alpha_{k,1}^1, \beta_{k,1}^1$. The dependence of the functionp$^{(1)}(\theta,t)$ on the coefficients can be represented in the following form:

We substitute the equation of contact pressure at the elementary friction site of the friction unit (9) into the dependence (7):

$$U = \sum_{i=1}^{M} \left\{ \varepsilon[x_1 f_1(\theta_i) + x_2 f_2(\theta_i) + \cdots + x_{4m+2} f_{4m+2}(\theta_i)] + p^{(0)}(\theta_i, t) - \overline{p} \right\}^2 \rightarrow \min \qquad (10)$$

We define the partial derivatives with respect to each unknown parameter and equate them to zero according to the principle of the least squares method:

$$\begin{aligned}
\frac{1}{2}\frac{\partial U}{\partial x_1} &= \sum_{i=1}^{M} f_1(\theta_i) \left\{ \varepsilon[x_1 f_1(\theta_i) + x_2 f_2(\theta_i) + \ldots + x_{4m+2} f_{4m+2}(\theta_i)] + p^{(0)}(\theta_i, t) - \overline{p} \right\} = 0; \\
\frac{1}{2}\frac{\partial U}{\partial x_2} &= \sum_{i=1}^{M} f_2(\theta_i) \left\{ \varepsilon[x_1 f_1(\theta_i) + x_2 f_2(\theta_i) + \ldots + x_{4m+2} f_{4m+2}(\theta_i)] + p^{(0)}(\theta_i, t) - \overline{p} \right\} = 0; \\
\frac{1}{2}\frac{\partial U}{\partial x_{4m+2}} &= \sum_{i=1}^{M} f_{4m+2}(\theta_i) \left\{ \varepsilon[x_1 f_1(\theta_i) + x_2 f_2(\theta_i) + \ldots + x_{4m+2} f_{4m+2}(\theta_i)] + p^{(0)}(\theta_i, t) - \overline{p} \right\} = 0; \\
\frac{1}{2}\frac{\partial U}{\partial \overline{p}} &= \sum_{i=1}^{M} (-1) \left\{ \varepsilon[x_1 f_1(\theta_i) + x_2 f_2(\theta_i) + \ldots + x_{4m+2} f_{4m+2}(\theta_i)] + p^{(0)}(\theta_i, t) - \overline{p} \right\} = 0;
\end{aligned} \qquad (11)$$

The coefficients for unknowns in the system of equations (11) depend on the interaction time of the contacting pads, the specific loads on the friction pairs, and the temperature developed on the contacting surfaces of the friction unit. For design, the solution of greatest interest is the initial interaction of the contacting friction pairs, i.e. t = 0.

Solving the system of linear equations (11) for a fixed instant of time by the Gauss method, using computer simulation as applied to disk brake mechanisms, we find the desired coefficients of the Fourier series to solve the contact problem.

The results of computer simulation of the coefficients of the contact problem at the beginning of the interaction of a friction pair (t = 0) are given in table. 1.

To find the contact pressure p, it is necessary to set the contact time of the working platforms (t = 15s). Simulations of the contact pressure of an elementary site will be considered for two cases. In the first, the polar angle of the contact pad of the friction unit will be 30 degrees, in the second, the polar angle increases to 50 degrees. In addition to the polar angle of interaction of the elementary areas of the contact pair of the friction unit, the law of the technological roughnesses of the area will change.

Table 1. Calculation results

$\alpha_{k,0}^1$	$\beta_{k,0}^1$	$\alpha_{k,1}^1$	$\beta_{k,1}^1$	$\alpha_{k,2}^1$	$\beta_{k,2}^1$	$\alpha_{k,3}^1$	$\beta_{k,3}^1$
0,379	0,411	0,323	0,231	0,128	0,455	0,279	0,195
$\alpha_1^{(1)}$	$\alpha_2^{(1)}$	$\alpha_{32}^{(1)}$	$\beta_1^{(1)}$	$\beta_2^{(1)}$	$\beta_3^{(1)}$		
0,582	0,779	0,976	0,388	0,652	0,849		

The results of modeling the contact pressure of the elementary site of the friction unit versus the value of the polar angle are presented in Figure 2. Curves 1 and 3 correspond to the cosine law of the distribution of technological roughnesses, curves 2 and 4 of the sinusoidal distribution of technological roughness. The knowledge of the contact pressure of the working platform of the friction unit allows one to further find the temperature distribution on the surfaces of the friction unit, to determine the stress-strain state, as well as their wear.

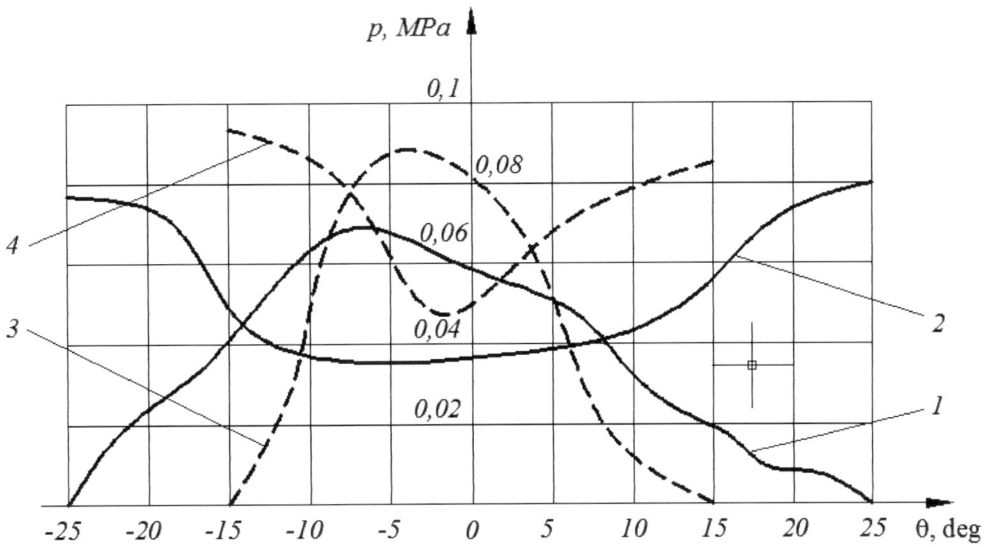

Figure 2. Dependences of the contact pressure of the elementary pad of the contact pair of the friction unit on the value of the polar angle: curves 1 and 2, the angle of the contact pad 50 degrees, curves 3 and 4, the angle of the contact pad 30 degrees.

3. Conclusion

Modeling showed that under the same braking forces and the values of the free parameters of the friction unit parts, the contact pressure of the elementary area depends both on the nano-geometry of the surfaces of the working surfaces of the friction unit and on the roughness class. A higher roughness class helps to reduce the contact pressure of the working surfaces of the friction unit and, accordingly, the temperature on the working surfaces of its elements. This fact is explained by the fact that the energy for deformation of the protrusions of the nano-geometry of the contact surface is less affected, and thus the working surfaces of the contact pads absorb less energy. According to the sinusoidal and cosinosidal laws of the distribution of technological roughnesses, the maximum contact pressure will be shifted in the direction of rotation of the metal friction element relative to the axis of symmetry of the elementary area. The simulations also confirmed with decreasing girth, contact pressure will increase.

The developed mathematical model of the contact pressure of the working areas of the friction unit allows determining the dependence of the nano-geometrical working surfaces of the contact areas of the friction unit. These studies allow designing friction units taking into account the distribution of contact pressure of the working pairs of the friction unit, which will affect the stress-strain and thermal state of the working pairs of the friction unit.

References

[1] Volchenko N A, Volchenko D A, Polyakov P A, Fedotov E S and Evchenko A S 2019 Pulse-contact frictional interaction of microprotrusions of friction pairs of brake *devices IOP Conf. Series: Materials Science and Engineering* **560**

[2] Kolesnikov I V 2003 *Thermophysical processes in metal-polymer tribosystems* (Moscow: Nauka)

[3] Volchenko N A, Krasin P S, Volchenko D A and Voznyi A V 2018 Electrochemistry with double electrical layers in frictional interaction metal-polymer tribolink *IOP Conference Series: Materials Science and Engineering* **327**

[4] Kolesnikov I V 2014 Triboelectric phenomena at metal-polymer friction contact and their dependence on temperature *Engineering journal of Don* **4-1 (31)**

[5] Kindrachuk M V, Vol'chenko D A, Vol'chenko N A, Stebeletskaya N M and Voznyi A V 2017 Influence of Hydrogen on the Wear Resistance of Materials in the Friction Couples of Braking Units *Materials Science* **53 (2)** 282-288
[6] Kindrachuk M V, Vol'chenko A I, Vol'chenko D A, Zhuravlev D Yu and Chufus V M 2019 Electrodynamics of the Thermal Contact Friction Interaction in Metal-Polymer Friction Couples *Ukrainian journal Fizyko-Khimichna Mekhanika Materialiv* **54 (1)** 69–77
[7] Kindrachuk M, Volchenko A, Volchenko D, Volchenko N, Poliakov P, Tisov O and Kornienko A 2019 Polymeres with enhanced energy capacity modified by semiconductor materials *Journal "Functional Materials"* **26-3** 629-634
[8] Volchenko N A and Polyakov P A 2012 Nanotribology in the interaction of the surface layers of friction pairs of drum-shoe brakes *Vestnik Saratov State Technical University* **1 - 1 (63)** 41-47
[9] Kolesnikov I V, Chebakov M I and Danilchenko S A 2019 Analysis of the effect of the hardened surface layer properties of the tribocontact parts on their wear and stress-strain state *Advanced Materials. Springer Proceedings in Physics* **224** 637-642
[10] Kolesnikov I V, Chebakov M I and Danilchenko S A 2019 The Nonstationary Contact Problem of Thermoelasticity for a Ring in the Presence of a Reinforcing Layer on the Friction Surface *Journal of Friction and Wear* **4** 447-455

IOP Conf. Series: Materials Science and Engineering **843** (2020) 012002 doi:10.1088/1757-899X/843/1/012002

Cooperation of future automated machinery specialists in informational context as a means of communication during vocational training process

N Y Saigushev, O A Vedeneeva, Yu B Melekhova and A A Tsaran

Nosov Magnitogorsk State Technical University, 38, Lenina ave., Magnitogorsk, 455000, Russia

email: nikolay74rus@mail.ru

Abstract. The article proves the significance of the issue of academic interaction and cybersecurity in information society. Subject-to-subject relations are examined as a form of relationship humanization when using computer technologies during the vocational training of future automated machinery specialists, which results in their personal advancement and transformation. It is necessary to take into account that in the context of social development over the last years we all witness the aggravating problem with the security of computers, mobile phones and other different devices representing items that are most often subjected to the attack. In order to resist the new kind of attacks effectively every member of information society must possess a certain minimum level of knowledge, be familiar with the respective "cybersecurity culture" and willing to participate in the active struggle for the freedom of information and communication technologies from various cyber fraudsters, criminals, etc.

1. Introduction

Academic interaction process is diverse and addresses the problems of personality formation and development, gaining social experience and enhancement of both the teacher and the student. At the same time, there is a main attribute that can be used as a basis to clearly distinguish between the types of interaction and build their models. This attribute constitutes the discrepancy between the goals and objectives set by a higher education institution and those pursued by students at this given moment.

We support the position of V.G. Maralov [1] who believes that academic interaction involves such an interplay between teachers and students that results in their personal advancement and transformation.

2. Materials and methods

Methodological foundation of the study is based on the vocational training theory, reflexive activity approach, system approach, modular training theory, contextual education training, society informatization theory, theory of computer technologies [2,3,4,5,6,7]. The issues of cybersecurity culture development, software classification for educational purposes and academic requirements for tool funds were addressed by S.M. Kalanova, V.P. Linkova, G.A. Lisiyev, S.V. Panyukova, M.I. Potev and others [8,9].

Subject, hypothesis and objectives of the study determined the selected set of research methods including theoretical analysis; examination and generalization of advanced academic experience;

academic observation; comprehensive diagnostics using questionnaires and testing; conversations with students and teachers; monitoring; experiment; description; analysis; generalization and systematization of emirical data; methods of electronic data processing and visual representation of findings and simulation, methods of mathematical statistics, qualitative and quantitative analysis of results, mathematical and statistical data processing, analysis and generalization of theoretical conclusions.

3. Results and Discussion

Being the constituents of this trend, personal needs and interests – in our case, these include advanced training and self-organization of a mining production specialist's personality - acquire personal value for him through realizing his mission.

We shall turn our attention to interaction based on subject-to-subject relations where the principle of parity (equality) is realized.

Pedagogics that is built on the system of subject-to-subject relations is referred to as cooperative pedagogics, polysubjective approach, person-centered model of interaction and pedagogics of non-violence. Here, both a teacher and a student are equally recognized as subjects of pedagogical process and have a certain level of freedom in organizing their activities, which is characterized by the ability to make a choice and develop their own personality through that. The main discrepancy is addressed by means of cooperation rather than enforcement. Therefore, such personal qualities as the ability for self-actualization and creative development, capability to create relations based on mutual acceptance and understanding as well as active approach and initiativity as the form of its expression take on particular importance. That way both the teacher and the student acquire the right for individuality.

Special research and practices based on subject-to-subject relations showed that the democratic management style was prevailing and the consistently positive type of relationship was central in this case.

The communication methods including personality apprehension, recognition and acceptance are based on the teacher's developing capability to accept the student's position, consider his point of view and not to ignore his feelings and emotions.

Cooperation is the tactics of communication. An example of such cooperation are the succesful practices of using information and communication technologies during the vocational training of future automated machinery specialists.

Using a computer and information technologies in class can address the following tasks:

- educational: studying the computer as an object of cognition; using the computer and application software in academic and professional activities rationally, competently and effectively;
- pedagogical: mastering the studied course content in a quick and qualitative way; visualizing the content; providing for individual learning paths for students;
- organizational: computer testing, record-keeping and planning.

As demonstrated by empirical practice with students, the formation of relationship humanization will be effective if the use of computer technologies during the interaction involved with vocational training of future automated machinery specialists is subject-to-subject.

Therefore, the above represents the model built on the system of subject-to-subject relations and creates favorable conditions to address the main discrepancy between the goals and objectives set by the teacher and those defined by the student.

As proved by the experience gained in the course of activities involving future automated machinery specialists, the following applications of computer technologies in vocational student training had strong potential for relationship humanization:

- studying the informatics module of the "Information science and information and communication technologies" course;
- computer-aided classes of the course;
- computer testing;
- referring to the resources of the educational institution's information center;

- visiting the educational institution's website;
- Internet activities;
- organizing and conducting extracurricular activities;
- publicating topical bulletins;
- publicating wall newspapers and designing stands;
- management (keeping an electronic journal);
- preparing documentation.

The computer allows changing the control over the activities of future automated machinery specialists in a quality manner and ensure flexible management of academic and professional activities by doing so. The teacher's role is equally important here. He selects software for the class along with the learning material and individual tasks, assists students in the course of their activities and evaluates their knowledge and development. The application of computer equipment makes and allows choosing the optimum form of training. Personality traits of both the student and the teacher are developed within the scope of this model and the activities take the form of independent action and enhanced by own personal experience.

When using a computer, verbal communicative activities can be regarded in terms of three aspects:

- firstly, as casual online communication between students by using electronic mail and information networks, i.e. as an authentic written dialogue between communication partners;
- secondly, as an interactive conversation between future automated machinery specialists and the computer, which pursues the real goals of communication, i.e. as a man-machine dialogue;
- thirdly, as communication between students during their work with education software that functions as a means of simulating the conditions of communicative situation.

Yet, it is necessary to take into account that in the context of social development over the last years we all witness the aggravating problem with the security of computers, mobile phones and other different devices representing items that are most often subjected to the attack. We can state without exaggeration that the intensive expansion in the number of the global Internet subscribers implies an increased number of Internet attacks ("cyberattacks") and the use of the modern personal computer provides the attackers with a tool for spying and different destructive activities, which is unique in terms of its features. New related terms such as cybercrime, cyberterrorism and cybersecurity have come into existence these days.

Cyberattacks involve unintended or unauthorized access to, using, handling and interrupting or destroying through the use of electronic means of information or electronic devices, processes and physical infrastructure used for information processing [10]. We can see that cyberattacks can cause serious, and sometimes also irrecoverable, damage to the society within informational context.

In order to resist the new kind of attacks effectively every member of information society must possess a certain minimum level of knowledge, be familiar with the respective "cybersecurity culture" and willing to participate in the active struggle for the freedom of information and communication technologies from various cyber fraudsters, criminals, etc.

Definition of the term "cybersecurity" states that it refers to the conditions of protection against physical, mental, financial, political, emotional, professional, psychological, educational or other types of influences existing in the cyberspace, which could be considered undesired [11]. In other words, cybersecurity refers to protection against intentional and targeted cyberattacks.

When analyzing the general definition of the term "culture", it refers to a set of knowledge, beliefs and behaviors based on symbolic thinking and social learning [12,13].

Therefore, by the term "cybersecurity culture" we understand a dedicated set of knowledge and behaviors focused on the protection against intentional and targeted cyberattacks within informational context.

According to statistics and social research, with the rapidly increasing level of using global information and communication networks (for example, in 2017 the number of Internet users in Russia reached 70 million people) the maturity of cybersecurity culture appears to be at an extremely low level (only about 10 per cent of Internet users in Russia have at least the minimum idea about the threats that they are subjected to themselves or expose others to while working with the Internet).

The number of various threats existing in the Internet these days is growing. At the same time, the number of those subjected to the attacks, which includes teenagers, students and schoolchildren, is not getting lower.

Due to the increasing need for creating the conditions to develop the cybersecurity culture, the Concept of the cybersecurity strategy in the Russian Federation was drafted, which included the following:

- providing for the improvement in the skill level of Russian citizens with regard to ensuring cybersecurity by way of developing new and expanding the existing educational programs and organizing awareness-raising campaigns;
- suggesting improvement, approval and implementation of educational standards for specialist training and retraining relating to cybersecurity;
- ensuring the tasks of developing and introducing the course on information security into the learning process offered by educational organizations of various levels, which includes modules on providing cybersecurity or supplementing the existing courses with the above modules;
- arranging for the organization of measures on stimulating public-private partnership with regard to supplementary vocational education in the field of cybersecurity [14,15,16,17].

Based on the above mentioned, we can see that the government is certainly interested in developing cybersecurity culture in the society overall and particularly among students.

Efforts are made today to teach this culture both at secondary schools and higher education institutions. However, according to the practical experience, this constitutes just a small part of what is necessary for the development of cybersecurity culture in information-oriented society. The students' vision of cybersecurity is not enough developed due to the lack of targeted systematic measures on cybersecurity culture development in the students' educational environment.

The main goal of the methodological service is to ensure accessibility and improve the quality of educational services through the modernization of education, aimed at ensuring compliance of state educational standards with available financial resources [18,19,20,21]. At all levels of the educational system, the improvement of the quality and innovative nature of education should be ensured by:

- introduction of new educational technologies,
- development of interactive forms of learning,
- wide use of design methods and methods that allow simulating real situations,
- modern training programs.

4. Conclusion

The following approaches can be proposed to address the issue of using computer technologies during the vocational training of future automated machinery specialists:

- solving professional and academic tasks that promote skills and abilities related to safe and rational behavior among future automated machinery specialists when working with software and the Internet;
- developing a dedicated practice-oriented course for future automated machinery specialists, which will not only introduce them to each type of cyberthreat in detail but also suggest the elaboration of correct actions to be implemented when facing various cyberattacks.

References

[1] Maralov V G 1992 *The psychology of academic interaction between the supervisor and children* (Moscow)
[2] Batyshev S Ya 1984 *Production pedagogy* (Moscow)
[3] Vygotsky L S 1986 *Psychology of Art* (Moscow: Art)
[4] Bespalko V P 2006 *Theory of creation and application* (Moscow: Research Institute of School)
[5] Lavrentieva N B, Nechaeva A V 2010 *Educational ethics* (Barnaul: AltSTU Publishing House)
[6] Verbitsky A A 2017 *Higher education in Russia* **6** 57-67

IOP Conf. Series: Materials Science and Engineering **843** (2020) 012002 doi:10.1088/1757-899X/843/1/012002

[7] Gershunsky B S 1986 *Fundamentals of electronics and microelectronics* (Kiev)
[8] Kalanova Sh M 1998 *Personification of vocational training in the system of university education* (Almaty)
[9] Panyukova S V 2011 *Education management: theory and practice* **1 (1)** 126-165
[10] Grishin S E 2011 *Herald of the Saratov State Socio-Economic University* **4** 170 – 173
[11] Alpeyev A S 2014 *Matters of cybersecurity* **5 (8)** 39 – 42
[12] Saigushev N Y and Vedeneeva O A 2018 *Humanitarian and pedagogical research* **2(4)** 6-11
[13] Mikhailova U V, Saigushev N Ya, Vedeneeva O A and Tsaran A A 2018 *Journal of Physics: Conference Series*, **1015 (3)** 042054
[14] Saigushev N Y, Vedeneeva O A and Melekhova Y B 2018 *Journal of Physics: Conference Series* **1015 (3)** 032119
[15] Orinina L, Kashuba I and Samarokova I 2019 *Perspectives of science and education* **1 (37)** 316-326
[16] Gnevek O, Savva L, Saigushev N, Vedeneeva O and Shepilova N 2017 *India* **97(21)** 195-217
[17] Yusupova Z, Shakurova M, Saygushev N, Vedeneyeva O and Kashina S 2016 *International review of management and marketing* **6(2)** 403-409
[18] Saigushev N Y, Vedeneeva O A, Kashuba I V, Tsaran A A and Shcherbakova I A 2019 *IOP Conf. Ser.: Mater. Sci. Eng.* **560** 012038
[19] Mikhailova U V, Saigushev N Ya, Vedeneeva O A and Tsaran A A 2018 *Journal of Physics: Conference Series*, **1015 (3)** 042054
[20] Ardashkin I, Martyushev N and Bezborodov V 2015 *Procedia - Social and Behavioral Sciences*, **166** 227-231
[21] Kameneva G A and Bondarenko T A 2018 *Novosibirsk State Pedagogical University Bulletin* **4** 172-186
[22] Gershunsky B S *Fundamentals of Electronics and Microelectronics* (Kiev)

IOP Conf. Series: Materials Science and Engineering **843** (2020) 012003 doi:10.1088/1757-899X/843/1/012003

Numerical simulation of pulsed calibration of welded tubular part

N V Kurlaev, N A Ryngach, F M Tagoev and M E Ahmed Soliman

Novosibirsk State Technical University, 20, Karl Marks Ave., Novosibirsk, 630073, Russia

E-mail: kurlaev@corp.nstu.ru

Abstract. The article presents the analysis of electrodynamic processes during magnetic-pulse "expansion" of thin-walled tubular parts. The authors consider the cases where tubular parts are made by welding. As a result of welding, a deviation from the round shape occurs which requires calibration. Particular attention is paid to the case when the welding zone of a metal thin-walled tube touches the die. For the dynamic analysis of large deformations, the finite element method (FEM) based on the ANSYS software package is used, whose feature is the ability to implement a comprehensive solution to this class of problems. Using numerical calculations and optimization for pulsed metal forming, in particular pressure pulsed magnetic field (PMF), a nomogram was constructed to determine the required technological regimes that take into account the properties of the material and the pressure amplitude of the pulsed magnetic field.

1. Introduction

Nowadays the important problem in aircraft construction is the creation of reliable and high-resource pipelines that operate under difficult loading conditions. The pipelines of modern aircraft are subjected to high pressures, hydraulic shocks and pulsating loads, so they have high demands on the mechanical properties of the material, the quality of the inner and outer surfaces, shape retention, and minimal wall thinning.

A large number of pipeline failures are associated with a decrease in wall thickness in places of bending, transition from one diameter to another and joints. Significantly reduce the efficiency of such factors as the undulation of the walls and excessive ellipse, phenomena, accompanying the processes of manufacturing pipelines [1].

The actual problem is how to minimize production costs and at the same time ensure the specified accuracy of geometric parameters, the specified mechanical properties, the required surface roughness and performance characteristics of parts, assemblies and systems for applications in aircraft structures, modern metals and alloys.

In this regard, one of the important tasks in the manufacture of aircraft is the development of new and improvement of existing technologies.

Currently, one of the most promising technologies is the technology of electromagnetic forming (EMF) process for metals, based on the use of large electrodynamic forces arising in conductive materials during the interaction of an external pulsed magnetic field (PMF) with eddy currents induced in the material [2, 10]. These are typical high-speed forming technologies that have various advantages compared to quasistatic forming, such as single-sided stamping, improved sheet metal forming and high precision forming.

2. Statement of the problem

In the manufacture of pipelines, welded fittings of sheet material are used [1]. As a result of welding, a deviation from the round shape occurs, which requires calibration. For aluminum alloys, rubber-pad stamping can be used for calibration, but for stainless steels like 08X22H6T, 08X21H6M2T, 08X18Г8H2T or titanium alloys there will be significant springing.

In this work, we study the pulsed pressure treatment of thin-walled welded metal pipes when the welding zone of a thin-walled pipe touches the die.

As a result of numerical simulation analysis using ANSYS finite element complex, in order to obtain a given shape without springing, it should be noted that the loading scheme during shaping should have the form, as shown in Figure 1. The pressure P is evenly distributed on the inner surface of the tubular part.

It is necessary to set the optimal pressure amplitude P_0, at which the tubular part is formed along the die, without springing and changing the wall thickness. By finding the minimum required pressure, the reduction in wall thickness is minimized.

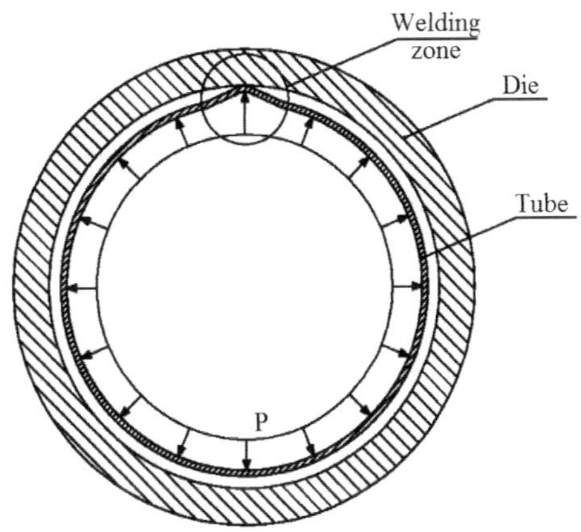

Figure 1. Scheme of loading of a tubular part during magnetic pulse calibration

3. Magnetic pulse stamping of tubular parts

The EMF process is mainly suitable for the deformation of highly conductive metal materials such as aluminum, copper and copper alloys. EMF is also an energy-saving and environmentally friendly production technology. The ultimate shaping of aluminum alloy material can be significantly enhanced by using EMF. EMF requires only one shape (for example, a die) for forming tubular workpieces with high accuracy. The values of the pressure amplitude PMF and pulse duration are determined based on the electrical parameters of the EMF process [3, 8]. The current flowing in the inductor when the capacitor Bank is discharged, approximately expressed by the harmonic oscillation equation:

$$I = \frac{U}{L\omega} \exp\left(-\frac{R}{2L} t\right) \sin\omega t \tag{1}$$

For magnetic-pulse "expansion" tubular parts, the pressure of PMF (P) on the fixed workpiece is approximated by a decaying quadratic sinusoid and the coefficient of pressure attenuation of PMF is introduced depending on the gap between the inductor and the workpiece [4] as shown in Figure 2:

$$P = P_0 \frac{\Delta}{\Delta + \mu} \exp(-\beta t) \sin^2(\omega t) \tag{2}$$

Where P_0 represents the conditional pressure amplitude, Δ represents the gap between the inductor and the workpiece, μ represents removing the workpiece from its original position, ω represents circular frequency of the discharge current and β represents attenuation coefficient.

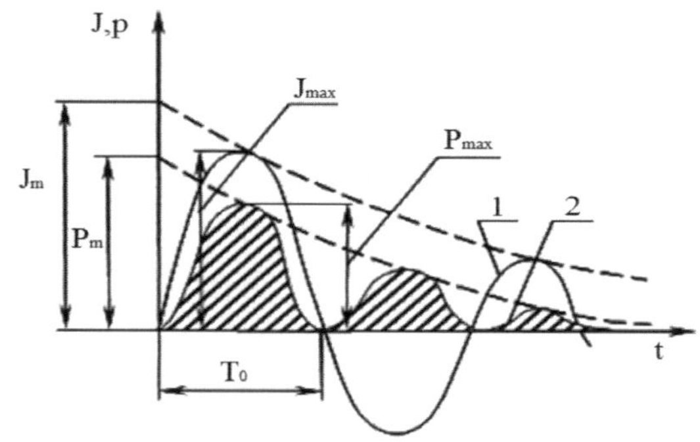

Figure 2. Dependence of the measurement of the discharge current J and pressure PMF P in time: 1 - current curve; 2 - pressure curve.

When analyzing the works devoted to the issues of numerical modeling using the ANSYS finite element complex given in [2, 7, 9], we can conclude that since the duration of the current discharge is much shorter than the duration of the workpiece deformation process, it is possible to neglect the influence of the movement of the workpiece during the discharge on the discharge parameters.

In this work, numerical modeling was carried out to study the dynamic behavior of a tubular part during pulse calibration.

As it is seen from Figure 1, the general EMF scheme in the expansion process where P represents the amplitude of the pulse loads. As a result, we obtain the deformation of the tubular part using a finite element model by using the influence of electromagnetic force.

4. Mathematical statement of the problem (choice of mathematical model)

Finite element analysis is one of the main methods of EMF analysis. In this work, we use the ANSYS software. The program solves a number of differential equations for deriving pressure, velocity, and deformation at different points in time [5, 6]. When choosing a mathematical model, it should be noted that the choice should be based on what result you want to get. The most effective is called such a mathematical model that provides a reliable solution (for example, with an acceptable error) with the least amount of time and computing power. At the same time, the solution using the FEM does not provide more information than is laid down in the solved mathematical model. For example, if the model does not include nonlinear properties of the material initially, the body will not begin to behave nonlinearly in the process of solving the problem. At this stage, the selection of equations describing the process under study is carried out. The analysis of processes is carried out on the basis of numerical modeling in the framework of continuum mechanics [7]. We use equations that are a consequence of the laws of conservation of mass, momentum, and energy. The system is closed by the relations of the elastoplastic behavior of the Prandtl-Reiss material. The basic equations have the following form as follows:

4.1. The equation of the trajectory of material particles:

$$\dot{x}_i = v_i \tag{3}$$

4.2. The continuity equation:

$$V_0\rho_0 = V\rho \tag{4}$$

4.3. The law of changing the momentum of a material particle:

$$\rho \dot{v}_i = \sigma_{ij,j} \quad (5)$$

4.4. Change in internal energy:

$$\rho \dot{e} = \sigma_{ij} \dot{\mathcal{E}}_{ij} \quad (6)$$

4.5. Strain rate tensor:

$$\dot{\mathcal{E}}_{ij} = 0.5(v_{i,j} + v_{j,i} \quad (7)$$

4.6. The components of the stress tensor are represented as follows:

$$\sigma_{ij} = -P\delta_{ij} + s_{ij} \quad (8)$$

where s_{ij} represents the components of the stress tensor deviator, which is responsible for the resistance to the shape change of the material particle and δ_{ij} represents Kronecker symbol. In the above equations (3-8), the following notation is used: each of the indices *i, j* sequentially takes the value (1, 2, 3); summation is performed over a repeating index, a dot over the symbol represents the time derivative along the trajectory; index after the decimal point represents the derivative with respect to the corresponding coordinate; x_i, v_i represent the components of the position and velocity vectors of the material particle, ρ_0, ρ represent the initial and current density of the medium, μ represnts shear modulus and P represnts the pressure.

4.7. The equation of the elastoplastic flow process is written in the Prandtl-Reiss form:

$$\dot{s}_{ij} + d\lambda' s_{ij} = 2\mu\left(\dot{\mathcal{E}}_{ij} - \frac{\dot{\mathcal{E}}_{ij}\delta_{ij}}{3}\right) \quad (9)$$

4.8. Under the condition of Guber-Mises plasticity:

$$s_{ij}s_{ij} \leq \frac{2}{3}Y_0^2 \quad (10)$$

where Y_0 represents the dynamic yield strength.

The load on the workpiece (see Figure 1) is set in the form of a uniformly distributed pressure from the inside of the pipe. The pressure value in time is set according to the first half-cycle of the quadratic sinusoid of pressure (*equation* 2), graphically presented in Figure 2.

5. Calculation of the mechanical characteristics of the process

Calculation of the mechanical characteristics of the process is an important step of the method, which allows determining the specific features of the performed technological operation. In this case, both the general regularities of the operation known during quasistatic metal forming, and specific relations of dynamic deformation of metals are used.

In the calculation, such mechanical characteristics of the process are determined as the work of deformation of the workpiece, necessary for the manufacture of a given part; kinetic energy of the deformable section of the workpiece; the required value of the pressure PMF . When analyzing the calibration processes, the required value of the pressure PMF is determined.

When processing tubular workpieces, the amplitude value of the pressure PMF should be sufficient both for the formation of the part and for its calibration without springing. The amplitude value of the pressure for quasistatic deformation to the boundary of the die can be determined in accordance with the well-known Laplace's equation:

$$P = \frac{Y_0 \cdot \delta_0}{R_{min}} \quad (11)$$

where R_{min} represents the minimum radius of curvature of the workpiece profile and δ_0 represnts the thickness of the workpiece. As a result of the analysis, when calculating the calibration processes of

IOP Conf. Series: Materials Science and Engineering **843** (2020) 012003 doi:10.1088/1757-899X/843/1/012003

tubular parts, the value of the pressure amplitude of the PMF should be 2...3 times higher than the pressure value which determined by the equation (11).

6. Result of the simulation, tubular parts are deformed by the pressure PMF

To assess the possibility of implementing such an approach in relation to this problem, this process was modeled using the ANSYS/ Explicit Dynamics finite element complex, whose feature is the ability to implement a comprehensive solution to this class of problems. Figure 3 shows, as an example, a 3D model of this problem and the result of the deformation of the tubular part.

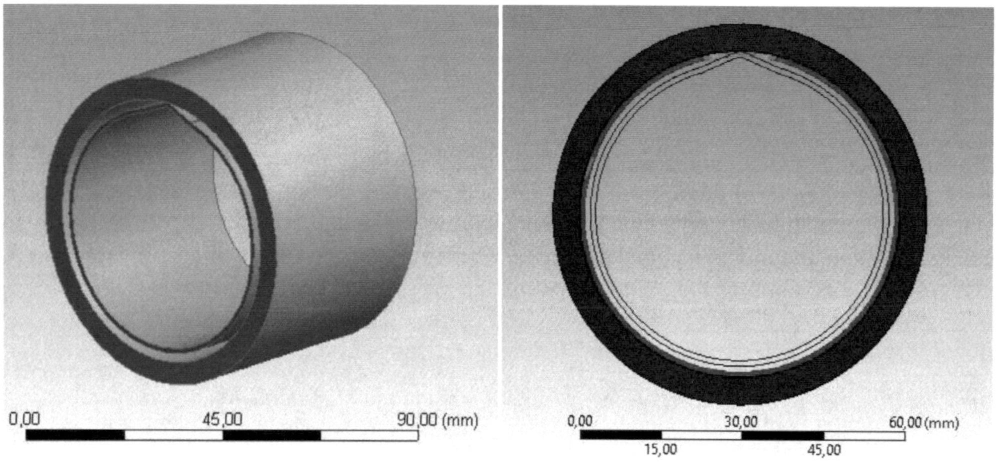

a b

Figure 3. 3D model of the tubular part: a – Initial state, b – Deformed state.

For the analysis, aluminum pipes made by welding were selected with different yield strength, thickness and diameter (as shown in table 1). As a die, a material was chosen whose yield strength exceeded the yield strength of the workpiece several times. The results of numerical modeling of loading processes of a tubular workpiece to obtain parts with a given accuracy were summarized in the form of nomograms. Nomograms for determining the optimal pressure (P) depending on the dynamic yield strength of the workpiece (Y_0) and the geometric parameters of the part (D_0, δ_0) are presented in Figure 4.

Table 1. Determination of the optimal pressure value (P, MPa) at the wall thickness of the tubular workpiece $\delta_0 = 1$ mm.

Y_0, (MPa)	D_0=50 (mm)	D_0=55 (mm)	D_0=60 (mm)
185	14	13	12
200	15	14	13
250	17	16	15
325	22	20	19

Thus, numerical modeling is necessary to study unobservable values (for example, behavior and strain rate during a process) in order to understand the physical phenomena. In addition, the simulation method can be useful in predicting experimental results that cannot be performed due to equipment limitations.

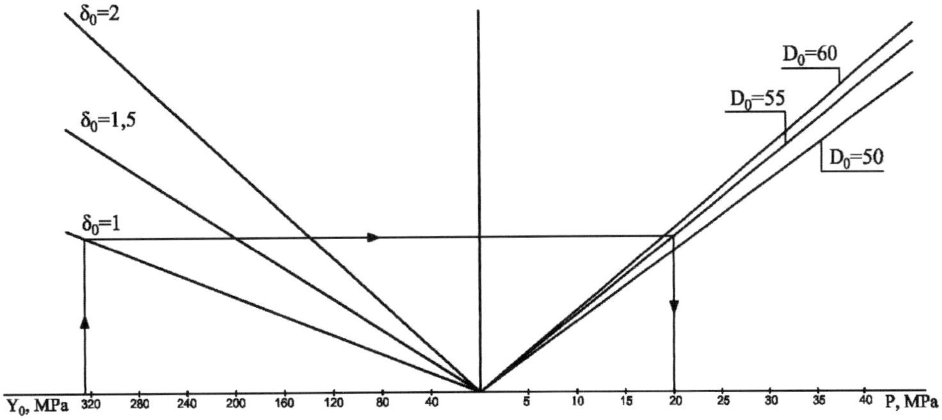

Figure 4. Nomogram for determining the optimal amplitude PMF (P) depending on the dynamic yield strength of the workpiece (Y_0) and the geometric parameters of the part (D_0, δ_0).

7. Conclusion

The conducted studies suggest the following:

- In the numerical simulation of pulsed pressure processing used for calibration tubular parts by manufactured welding, it was determined that the most effective method is to expand the workpiece into the covering die, where the pipe weld zone touches the die.
- As a result of the calculations of the calibration processes of tubular parts, the value of the pressure amplitude of the PMF should be 2...3 times higher than the value of the quasistatic deformation pressure.
- Based on the results of optimizing the parameters of magnetic pulse forming of tubular parts, a nomogram was constructed that allows finding the value of the pressure amplitude of the PMF for calibrating tubular parts with maximum accuracy.
- As a result of calculations, we can determine the optimal value of the PMF pressure for tubular parts with a dynamic yield strength from 120 to 330 MPa, a thickness from 1mm to 2mm, and a diameter from 50mm to 60mm.

Acknowledgments

The reported study was supported by the Department of Helicopter and Aircraft Construction, NSTU, Russian Federation, in accordance with order No. 144 of 01/30/2020. Project code included in the thematic plan of the Research Work for 2020 No. ТП-CBC-1_20.

References

[1] Feoktistov S I, Maryin B N, Maryin S B and Kolykhalov D G 2013 *Theory and Practice of Manufacturing Aircraft Pipeline Elements* (Komsomolsk-on-Amur: KnASTU Press)
[2] Glouschenkov VA 2012 *Special types of stamping, part 2, Dynamic methods of deformation, textbook* (Samara: SSAU)
[3] Belyy I V, Fertik S M and Khimenko L T 1977 *Handbook of magnetic pulse processing of metals* (Kharkiv: High school)
[4] Palamarchuk Yu E, Kirilin A N and Vershigorov V M 2008 Influence of inductor systems on the efficiency of magnetic pulse stamping, *Bulletin of Samara University. Aerospace engineering, technology and engineering* **1 (14)** 147-151
[5] Ivanov D V and Dol A V 2016 *Introduction to ANSYS WORKBENCH, a teaching aid for students*

of natural sciences (Saratov)

[6] Huei-Huang Lee 2019 *Finite Element Simulations with ANSYS Workbench 2019* (NCKU, Taiwan)

[7] Fomin F M, Gulidov A I, Sapozhnikov G A et al. 1999 *High-speed interaction between bodies* (Novosibirsk: Publishing House SB RAS)

[8] H Park, J Lee, J H Kim and D Kim 2019 Electromagnetic expansion joining between tubular and flat sheet component *Journal of Materials Processing Technology* **273** 116246

[9] H Yu, J Chen, W Liu, H Yin and Ch Li 2018 Electromagnetic forming of aluminum circular tubes into square tubes: Experiment and numerical *simulation Journal of Manufacturing Processes* **31** 613-623

[10] Psyk V, Risch D, Kinsey B L, Tekkaya A E and Kleiner M 2011 Electromagnetic forming—a review *Journal of Materials Processing Technology* **211** 787–829

IOP Conf. Series: Materials Science and Engineering **843** (2020) 012004 doi:10.1088/1757-899X/843/1/012004

Study of sides stamping under one-sided magnetic pulse loading on form blocks of various materials

K N Bobin, M B Detinov, M E Ahmed Soliman and N V Kurlaev

Novosibirsk State Technical University, 20, Karl Marks Ave., Novosibirsk, 630073, Russia

E-mail: kurlaev@corp.nstu.ru

Abstract. In this paper, the influence of the tooling material on the accuracy of forming the resulting aluminum alloy part under one-sided magnetic pulse loading was estimated. As the materials of the form blocks were used: dry pine, steel 45 and PLA-plastic. Numerical simulation was carried out using the ANSYS software package in the Explicit Dynamics module. Studies have shown the dependence of the nature deformation of the workpiece and the final geometry of the resulting part from the mechanical properties of the form block materials.

1. Introduction

Magnetic pulse stamping (MPS) of parts has been used since the end of the 50's of the last century, however, despite the fairly extensive experience in researching this process [1-8], the actual practical application of the technology in aircraft production has a number of practical difficulties [4, 9-13]. One of these problems is the choice of the optimal material of technological equipment for precise shaping of aluminum alloy profiles by a pulsed magnetic field (PMF). Obtaining high-quality geometry profiles of aircraft by magnetic pulse processing (MPP) is complicated by the asymmetry of the pulse loading, which imposes increased requirements for the choice of material of forming surfaces in comparison with conventional stamping conditions. The purpose of this work was to assess the impact of tooling on the accuracy of forming of the resulting aluminum alloy part in magnetic pulse stamping (MPS) processes with one-sided loading.

2. Methodology of theoretical research

To determine the effect of the tooling material on the accuracy of forming the resulting aluminum alloy part under one-sided magnetic pulse loading, a simulation of magnetic pulse stamping of the straight side of various form block materials was performed (shown in Figure 1) using the ANSYS software package in the Explicit Dynamics module. The calculation area consisted of three parts: the workpiece, the form block and the inductor. The thickness of the workpiece was 2 mm, and the height of the stamped side was 7 mm. The sizes of the form blocks were chosen deliberately large sizes of samples with a radius of rounding along the bending line equal to 3.2 mm. The lower surface of the form block and the upper surface of the inductor were restricted by absolutely rigid barriers. A restriction on waste was imposed on the left edge of the workpiece, which in practice corresponds to a more reliable fixation of the workpiece on the hairpin holes. When the surfaces of the workpiece and the form block, the workpiece and inductor come into contact, contact with friction is assumed.

Modeling of the inductor was carried out exclusively to impose restrictions on the significant waste of the workpiece from the form block. Each calculated area was divided into quadrangular cells with automatic meshing. The total number of finite elements of the model was 5494.

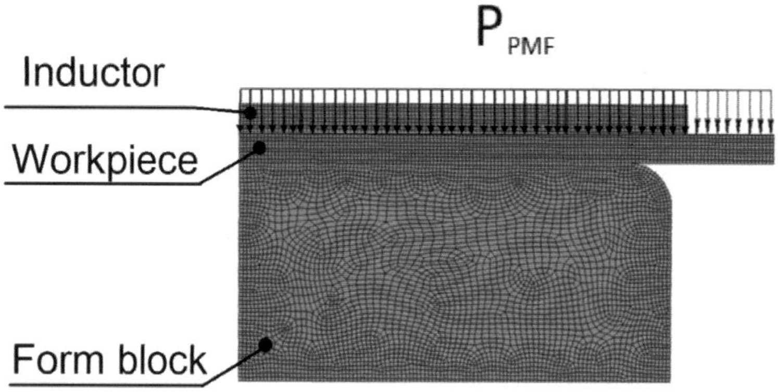

Figure 1. Scheme of loading by a pulsed magnetic field (PMF).

The mechanical properties of the workpiece material corresponded to the AMцM aluminum alloy [1]. As a model of the material used: bilinear-hardenable model. To account for the high-speed deformation of the material, instead of the static yield strength, a dynamic yield strength (Y_0) was used.

The dynamic yield strength of the AMsM alloy was estimated at the strain rates corresponding to magnetic pulse loading, was carried out using the method described in [14]. The estimated value was Y_0 = 120 MPa.

The mechanical properties of the workpiece material corresponded to: dry pine, steel 45 and PLA-plastic (PLA brand). All material specifications were taken from the ANSYS material library.

The calculation of the parameters of the discharge circuit of the magnetic pulse generator-inductor-workpiece system was carried out according to the methods described in [14, 15]. The estimated value of the frequency of the discharge circuit f was 20 kHz.

The pressure of the PMF (P_{PMF}) on the fixed workpiece was set by a decaying quadratic sinusoid [16] with only the first half-period of loading:

$$P_{PMF} = P_0 \sin^2(2\pi f t) \tag{1}$$

where P_0 represents pressure amplitude, f represents discharge current oscillation frequency.

The pressure attenuation of the PMF when moving away from the inductor was not taken into account, since the calculations used a sufficiently large value of the frequency of the discharge circuit in 20 kHz. The The amplitude of the magnetic pressure P_0 varied in the calculations from 10 to 150 MPa. The purpose of the numerical simulation was to determine the optimal pressure values of the PMF at which the most accurate part is manufactured with non-fit to the form block acceptable in aircraft manufacturing (less than 0.5 mm).

3. Numerical simulation results

Figure 2 shows the results of modeling the bending of the straight side by pressure of the PMF on a dry pine form block. With an increase in the PMF pressure, a predictable increase in the bending angle of the workpiece is observed (Figure 2 a, b). At a certain bending angle, due to the elastic properties of the material of the form block, the workpiece begins to spring from its radius part (Figure 2 c). A further increase in the pressure of the PMF does not allow increasing the fit of the side to the form block and begins first to lead to side lengthening and thinning of the material on the radius part of the workpiece (Figure 2 d, e), and then also increases the overall spring of the side and non-fit to the form block (Figure 2 f). At a pressure of about 140 MPa, the side is torn off on the radial part of the form block. Thus, the numerical simulation of the straight side bending by the pressure of the PMF on the dry pine form block shows the fundamental impossibility of obtaining parts with specified accuracy characteristics.

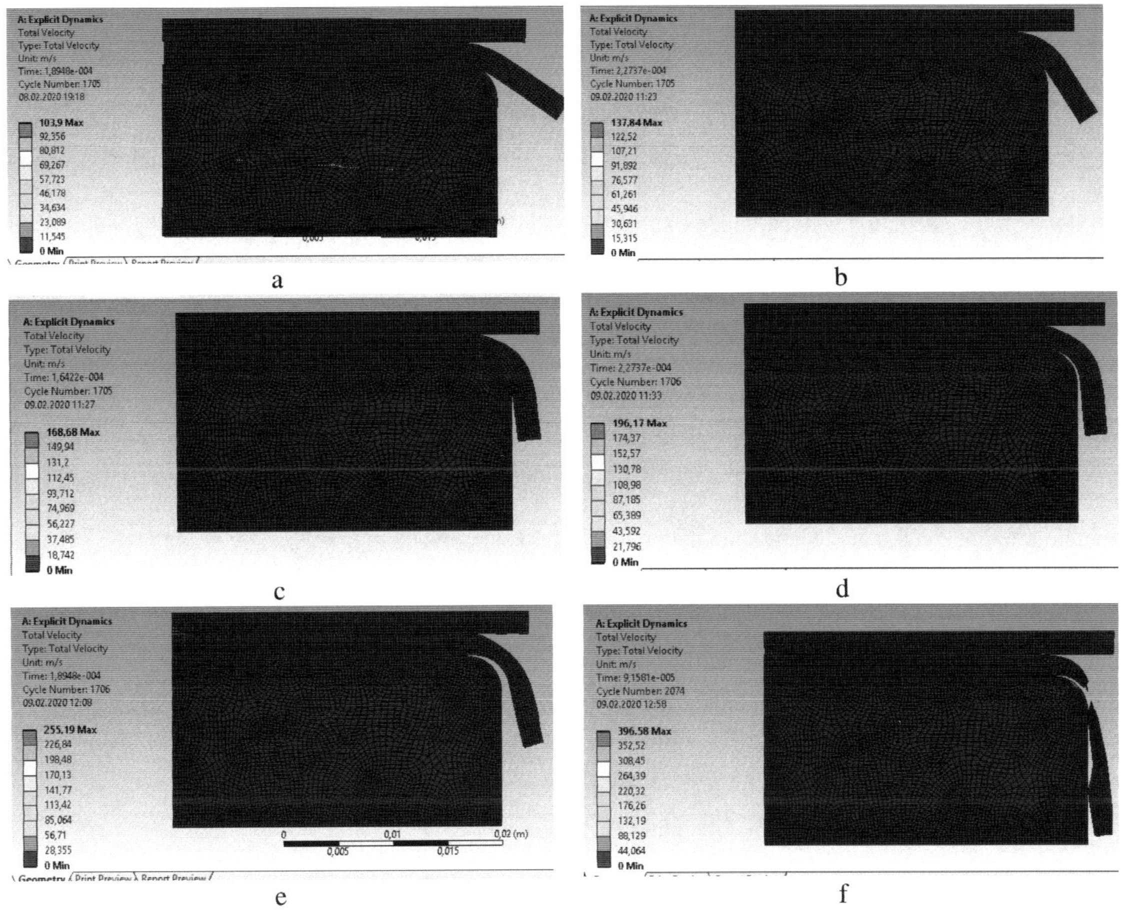

Figure 2. Kinematics of forming a workpiece on a dry pine form block: a – 30 MPa; b – 40 MPa; c – 50 MPa; d – 60 MPa; e – 70 MPa; f – 140 MPa.

Figure 3 shows the results of modeling the bending of the straight side by PMF pressure on the form block of steel 45. With an increase in the PMF pressure, there is also a predictable increase in the bending angle of the workpiece (Figure 3 a, b, c) and at an optimal pressure value of about 60 MPa, it is possible to obtain parts with minimal non-fit to the form block (Figure 3 e). A further increase in PMF pressure leads to side lengthening and thinning of the material on the radius part of the workpiece (Figure 3 d). At a pressure of about 125 MPa, the side is torn off on the radial part of the form block (Figure 3 f). Thus, the numerical simulation of bending the straight side of the PMF pressure on the form block of steel 45 shows the possibility of obtaining parts with maximum precision characteristics.

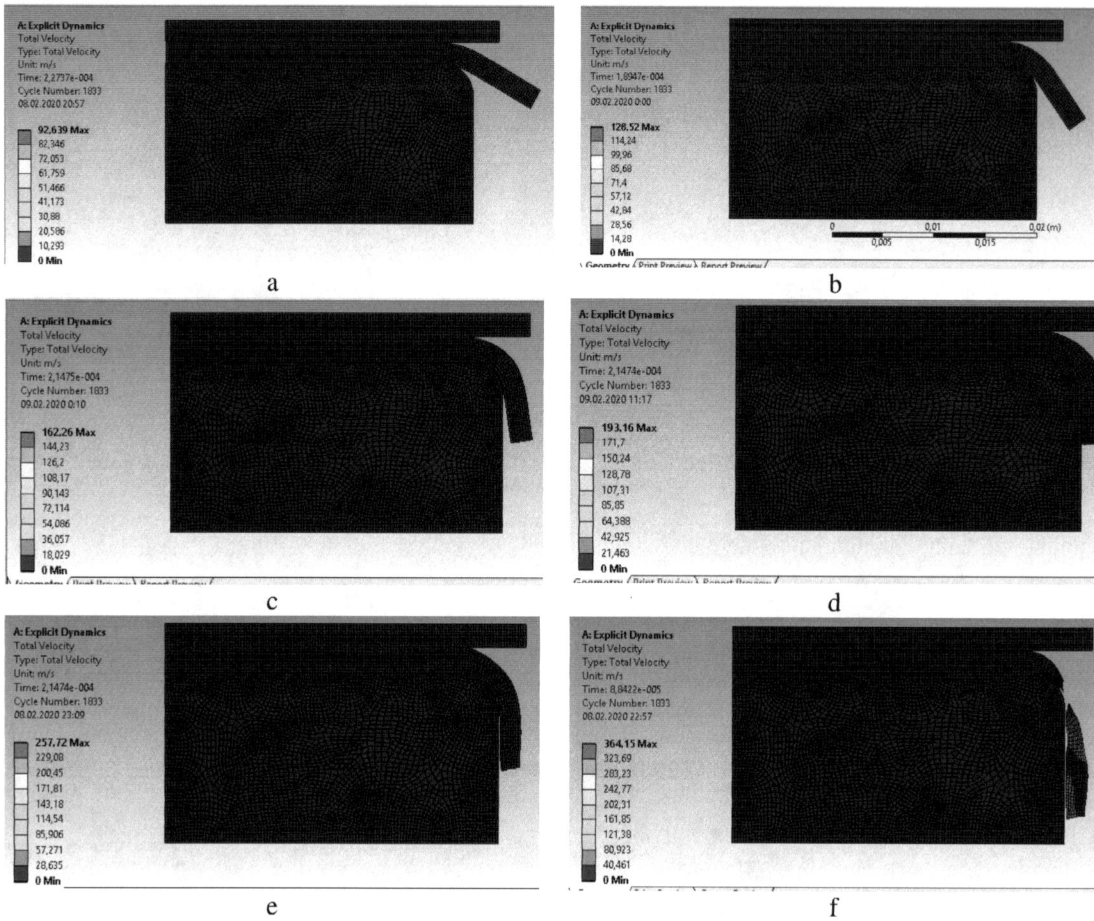

Figure 3. Kinematics of forming a workpiece on a steel 45 form block: a – 30 MPa; b – 40 MPa; c – 50 MPa; d – 60 MPa; e – 80 MPa; f – 125 MPa.

Figure 4 shows the results of a simulation of a straight side bending by PMF pressure on a PLA-plastic form block. The kinematics of the side forming on the PLA-plastic form block is similar to the kinematics on the dry pine form block (Figure 4 a, b, c), except for the possibility of obtaining a part with non-fit to the form block in the specified tolerance (Figure 4 c). Obviously, this is due to the lower stiffness of the plastic compared to dry pine. The workpiece crushes the radius part of the form block and due to this, the part is formed. In addition, in the process of deformation of the workpiece at the end of the impulse of applying PMF load, a significant "squatting" of the form block is observed (Figure 4 d). Obviously, in the actual manufacture of a part on such a form block, a gradual destruction of the tool will occur over time. With a further increase in pressure, as in the case of dry pine, the springing of the part increases and at a pressure of about 150 MPa, the side is torn off on the radial part of the form block (Figure 4 e). Thus, the numerical simulation of bending a straight side with the pressure of PMF on a PLA-plastic form block shows the possibility of obtaining parts with the required accuracy characteristics, but in practice this may be difficult due to the need for more accurate dosing of pulse energy and the possibility of destruction of the form block.

In general, in the transition from a less rigid material (PLA-plastic) to a more rigid material (steel 45), a decrease in the time of the forming process is noted due to the smaller contribution of the elastic characteristics of the tooling material to the value of the workpiece spring. With low stiffness of the material of the form block in the process of forming the workpiece, some fluctuations in the geometry of the tool are observed. These fluctuations in the geometry of the form block affect the workpiece in contact with it and cause an increase in the time of the forming process.

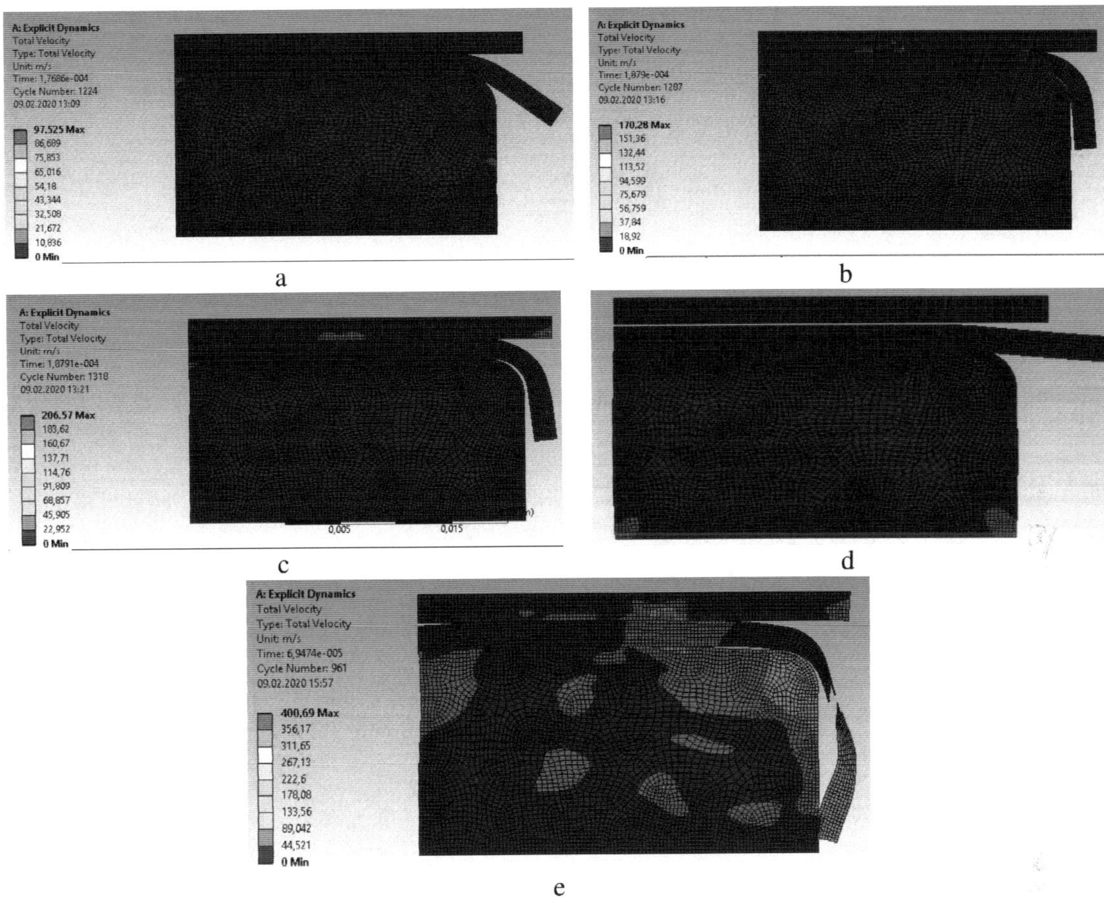

Figure 4. Kinematics of forming a workpiece on a PLA-plastic form block: a – 30 MPa; b – 50 MPa; c – 60 MPa; d – 50 MPa at time 2.2 μs; e – 145 MPa.

4. Conclusion

Thus, numerical studies using the finite element method of one-sided loading by a pulsed magnetic field of form blocks of various materials have shown:

- The most optimal material for use in form blocks with one-sided loading by a pulsed magnetic field is steel.
- With increasing rigidity of the materials of the form blocks in the process of flat magnetic pulse stamping, the value of springing of finished products decreases.
- When applying certain assumptions, numerical simulation of magnetic pulse processing can be performed in the ANSYS software package in the Explicit Dynamics module.
- Experimental studies are required to evaluate the accuracy and verify the validity of the numerical study performed using the Explicit Dynamics module of ANSYS program.

Acknowledgments

The reported study was supported by the Department of Helicopter and Aircraft Construction, NSTU, Russian Federation, in accordance with order No. 144 of 01/30/2020. Project code included in the thematic plan of the Research Work for 2020 No. ТП-СВС-1_20.

IOP Conf. Series: Materials Science and Engineering **843** (2020) 012004 doi:10.1088/1757-899X/843/1/012004

References

[1] Belyy I V, Fertik S M and Khimenko L T 1977 *Handbook of magnetic pulse processing of metals* (Kharkiv: High school)

[2] Talalaev A K 1992 *Inductors and systems for magnetic pulse processing of metals* (Moscow: Informtekhnika)

[3] Glouschenkov VA 1990 The main directions of development of magnetic pulse processing *Journal of Forging and Stamping production* **7** 11-13

[4] Galkin V P and Popov Yu A 2009 Analysis of the experience of mastering the latest pulsed electrophysical technologies in the USSR: on the example of magnetic pulse processing of metals and materials *Proceedings of the 3rd International Scientific and Technical Conference. In 2 volumes. Metallophysics, mechanics of materials, nanostructures and deformation processes. Samara* pp 62-67

[5] Baranov M I 2009 Progressive pulsed technologies for processing materials: history, physical foundations and technical capabilities *Journal of Electrical Engineering and Electromechanics* **1** 42-51

[6] Batygin Yu V, Serikov G S and Chaplygin E A 2006 Realization and prospects of magnetic-pulse methods in the development of advanced technologies of our time *Road transport, Kharkiv* 18 83-87

[7] Evandro Paese, Martin Geier, Roberto P Homrich, Pedro Rosa and Rodrigo Rossi 2019 Sheet metal electromagnetic forming using a flat spiral coil: Experiments, modeling, and validation *Journal of Materials Processing Technology* **263** 408-422

[8] Steven Woodward, Christian Weddeling, Glenn Daehn, Verena Psyk, Bill Carson and A. Erman Tekkaya 2011 Production of low-volume aviation components using disposable electromagnetic actuators *Journal of Materials Processing Technology* **211 (5)** 886-895

[9] Proskuryakov N E, Volodin AV and Lai D Z 2015 Calculation of energy efficiency of electromagnetic stamping processes of flat *workpiece Proceedings of the VII International Scientific and Practical Conference INNOVATION IN ENGINEERING (InMash-2015)* (Kemerovo: KSTU) 465-469

[10] Lebedev G M, Komarov A D, Isarovich G Z and Lysenko D N 1970 Inertial setting of corrugations during magnetic-pulse bending-forming of sheet parts of *aircraft Journal of Forging and Stamping production* **4** 25-28

[11] Volodin A V, Proskuryakov N E and Arkhangelsk N N 2017 Fundamentals of calculation of equipment and operations of electromagnetic flat stamping *Izvestya of Tula State University. Technical science* **11-1** 286-291

[12] Astapov V Yu, Khoroshko L L, Josdani M S and Khoroshko A L 2017 Production of sheet fringes of aircraft hatches using the magnetic pulse method and modeling using CAD systems *Proceedings of the Moscow Aviation Institute* **95**

[13] Fei Feng, Jianjun Li, Liang Huang, Xianlong Liu, Huizhuan Ma and Guodong Li 2017 The research of electromagnetic incremental forming process of Aluminum alloy sheet and the interaction between the die and the sheet *Procedia Engineering* **207** 329-334

[14] Karpukhin V F 2012 Determination of the parameters of magnetic pulse forming *Bulletin of the Samara State Aerospace University* **5 (36)** 228-232

[15] Belousov V S 1991 *Textbook of physical fundamentals of laser and magnetic pulse processing* (Novosibirsk: NETI)

[16] Yudaev V B 1995 *The use of pulsed magnetic fields for the shaping of aircraft parts: Textbook for coursework and design engineering diploma* (Moscow: MAI)

IOP Conf. Series: Materials Science and Engineering **843** (2020) 012005 doi:10.1088/1757-899X/843/1/012005

Environmental management in automated systems for engineering production

G S Zhetessova, V V Yurchenko, T Yu Nikonova, O M Zharkevich and N A Savelyeva

Karaganda State Technical University, 56, Nazarbaev ave., Karaganda, 100012, Kazakhstan

E-mail: nitka82@list.ru

Abstract. The article describes the issues of environmental management and environmental savings. The goal of environmental management is to conserve and save raw materials, materials, energy resources, and to improve the environmental situation. This article describes the technical aspects of environmental management in mechanical engineering production. Identification of environmental aspects occurs on the basis of the process approach. To manage aspects of environmental management in engineering production an automated program "Coating Mode Solver" has been developed. The program interface is described. Automated program selects the optimal technical modes of the process. Optimal process modes allow managing environmental aspects, reducing resource consumption and having a positive impact on the green economy.

1. Introduction

Successful actions to protect the environment and conserve natural resources constitute environmental management. More formally, environmental management allocates or develops resources; uses, restores, rehabilitates, controls or assesses resources [1].

Environmental management includes solutions, strategies, programs and projects [2].

Environmental management is an administrative function that develops, implements and controls the environmental policy of an enterprise. This can be viewed as an integrated mechanism that realizes the main goals of sustainable development [3].

Environmental management is broad in scope and includes all technical, economic and other aspects of the environment [4].

The main task of environmental management is to reduce the risk of environmental damage [5]. Globalization places high demands on resource management, therefore it is necessary to accelerate the development of management in order to improve efficiency and competitiveness [6].

As scientific and technological progress develops, the level of organization of the production process is improved, the processing of raw materials and materials is improved, incentives for saving material resources are introduced and the amount of waste per unit of production decreases [7]. A great influence on the amount of waste has a rationing of the consumption of material resources, that is, a scientifically based determination of the planned measure of production consumption of raw materials, materials, fuel and energy.

This process includes [8]:

- the study of factors affecting the development of norms of consumption of material resources for newly manufactured products or the implementation of a certain amount of work;
- adjustment and revision of current regulations;
- consideration and approval of new standards in the prescribed manner and their direct communication to work places.

At the same time, the tasks of using waste should be solved at the stage of the technological process development in manufacturing new products and the establishment of norms of material resources.

2. Methods and Prequisites

In engineering production, the technical aspects of environmental management are:
- prevention of accidental impacts on the environment;
- saving raw materials, the use of small energy resources;
- greater use of secondary resources;
- reducing waste, improving the quality of products for the environment;
- processing and use of waste in the production and consumption;
- reduction of situational risks with increased environmental and financial results in case of accidental impacts on the environment;
- reducing environmental risk for staff;
- improving the efficiency of industrial environmental monitoring and environmental monitoring of production processes (including monitoring the conservation and use of resources, sources of maximum release of pollutants into water bodies.

The environmental aspect is any element of production in an enterprise that is associated with environmental impact. The organization is required to identify controlled environmental aspects, it is assumed the possibility of further impact on them, in order to change [9]. Through environmental aspects determine the elements of the industrial activity of the enterprise, which have a negative impact on the environment and human health.

We use a process approach to the enterprise's activities in identifying environmental aspects (Figure 1).

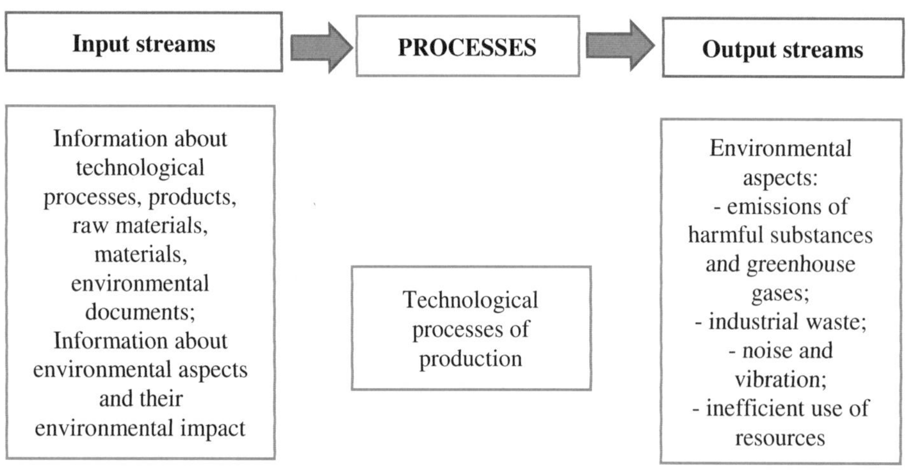

Figure 1. Process approach to the activities of the enterprise in the identification of environmental aspects.

Using the example of manufacturing rods of hydraulic racks for mechanized roof supports produced at LLP "KarGorMash-M", we will highlight aspects of environmental management that affect the ecological environment and human health (Table 1).

Table 1. Dependence of the sphericity coefficient of lead inclusions and average sizes of bronze lead inclusions on the modifier concentration

Environmental aspect	Aspect value	Main sources
The ability to spill waste lubricants, oils for roughing and finishing rods	3 l/year	Cutting machines work
Evaporation of chromic anhydride from electrolytic baths for plating surfaces rods	0.08 t/year	Electrolysis bath
Noise during chrome plating	80 dB	Electrolysis bath
Large power consumption	54 066 kW / year	Electrolysis bath
Abrasive dust	2.7 t/year	Cutting machines work
Metal dust	4.8 t/year	Cutting machines work

The management of environmental aspects in the production of rods for mechanized roof supports can be carried out through the use of computer-aided design of technological processes.

To reduce the impact of environmental aspects, it is proposed to optimize the technological process for the production of mechanized supports rods. Optimization is to eliminate roughing and finishing of the rod manufacture, as well as replacing the process of chromium plating on the electric arc spraying.

Electric arc spraying is a resource-saving technology in the production of mechanized supports rods. In addition to changing the production process, the computer program “Coating Mode Solver” was developed, which is intended for automated calculation of technological parameters for applying protective coatings on the surfaces of engineering objects - mechanized supports rods.

3. Results and Discussion

One of the ways out of the current situation is automation of technological preparation of production and, in particular, development and application of computer-aided design systems for production processes. Today there is a rather large number of such systems from various developers. The main disadvantage of such automated systems is that it is necessary to analyze and adjust results of designing of production processes according to a specific current situation at the place of production. The current situation is considerably connected with the fact that in the theory and methodology of process design there is a big number of recommendations on application of this or that technology solution which cannot be formalized and is applied only to specific cases. The vast majority of the existing CADs are constructed at best on the use of prototype solutions, and often the production engineers makes decisions themselves, based on the personal experience or experience developed by manufacturer. Formalization of process of designing and technology, which will be a basis of the system of design automation, will give a chance of production process design for various products, including new, not produced before. Thus, the creation of automation system of technological preparation of production integrating various on nature and content tasks of machine-building technology is an urgent task.

CodeGear RAD Studio Delphi 2009 was chosen as the development environment. It is a very powerful and convenient tool for developing applications for computers running the Windows operating system [10].

The development environment has a set of components and modules necessary for designing the external design of the forms, as well as the interface of the modules containing the source code, and corresponds to the ideology of object-oriented programming. The development environment contains a built-in code inspector, which improves the performance of the application development process.

Hardware:

- processor - at least Intel Pentium I 100 MHz;
- the amount of RAM - at least 256 MB;
- the amount of free hard disk space - at least 3 MB;
- availability of a backup or uninterrupted power supply.

Technical requirements:

- availability of an operating system at least Windows XP;
- screen area of at least 800x600 pixels;
- High Color palette at least 16 bits.

The choice of hardware was based on the minimum system requirements imposed by the selected software on the hardware.

When the "Coating Mode Solver.exe" file is launched (two mouse clicks), a program window appears on the screen (Figure 2), in which it is necessary to enter the parameters of the product being processed and select the material of the sprayed coating.

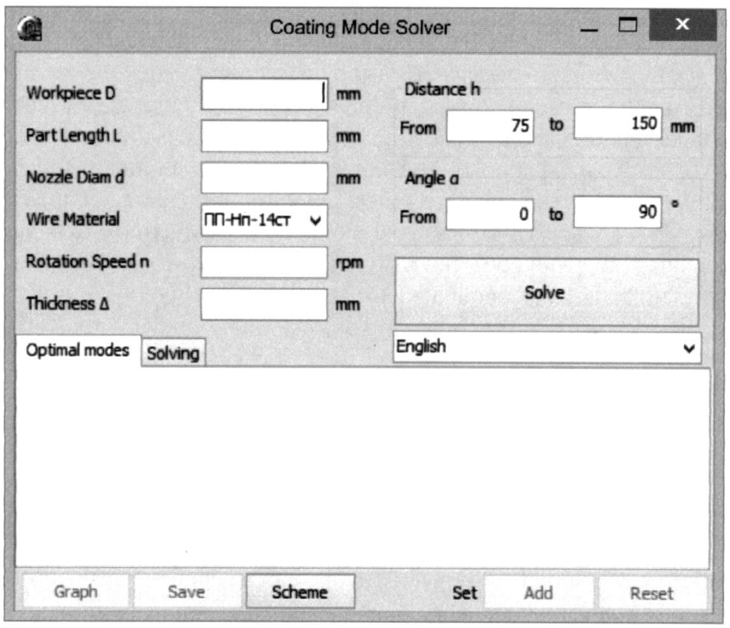

Figure 2. "Coating Mode Solver" program interface.

The proposed list of coating materials is not transferable, since foreign analogues of these materials have different mechanical and physical properties. Therefore, marking with foreign brands of steels is not provided as a translation. However, if it is necessary, these foreign steel grades can be added for selection in the program.

After filling in the proposed form with the data required for the calculation, it is necessary to click "Calculate" button.

After calculating the technological parameters of the coating, the optimal values of these parameters are automatically selected, at which the most accurate value of the coating thickness is achieved. In this case, the deviation error from the required value of the coating thickness does not exceed 2%. Results of the decision exceeding the given size of an error are excluded from calculation.

The history of the calculation can be viewed on the “Solving” tab, which displays the initial data entered, the results not exceeding 2% deviation errors obtained at each iteration, as well as the optimal calculation results.

The data displayed on the “Solve” tab is a program report and can be saved on a hard disk or portable device. To do this, click "Save" button on the bottom panel of the program.

The program provides the ability to compare the obtained parameters for each calculation. To perform the comparison, it is necessary to click “Add” button after each satisfying result operator to add the results to the set.

After adding the obtained parameters to the set, it is necessary to change any technological or geometrical parameters, for example, the diameter of the burner nozzle or the speed of part rotation, and carry out a new calculation and add the results to the set again. After the results of several calculations are added to the set, it is necessary to press the “Graph” button.

After performing this action, a window will open with graphs of dependencies obtained during the calculation. To change the displayed dependency graph, we change the value in the dependency field by selecting the necessary one from the list.

In order to save the image of the displayed graph, it is necessary to click “Save” button and select the path to save. To reset all added values, click “Reset” button on the bottom panel of the main application window. To view the design scheme, we click "Scheme" button in the bottom panel of the application. This program is easy to use for any user with basic computer literacy knowledge.

Using the program “Coating Mode Solver” gives more detailed data, reduces the time for writing the technological process, than the values of the same parameters set for reference. This program allows calculating the technological parameters of the deposition process for cylindrical parts of various dimensions and selecting the desired material for spraying from the list.

Thus, this program “Coating Mode Solver” calculates the technological parameters taking into account the initial data, and on the basis of these calculations provides optimal solutions and graphical dependencies, which is an important factor in the design of the technological process and in determining the environmental impact on the environment and human health. The program is an effective tool for resource conservation in machine-building enterprises.

4. Conclusions

1) Aspects of environmental management in the production of hydraulic rod props of mechanized roof supports were identified.
2) The authors developed software “Coating Mode Solve” for computer, designed for automated calculation of technological modes of applying a protective coating on the surface of engineering objects.
3) The selection of optimal modes in the “Coating Mode Solve” program allows reducing energy consumption by 3 times.
4) Optimum regimes of applying protective coatings in the production of cylindrical parts eliminate such environmental aspects as abrasive dust, metal dust.

References

[1] Maureen G 2013 *The Canadian Encyclopedia*
[2] Staib R 2005 *Environmental Management and Decision Making for Business* (New York, Palgrave Macmillan)
[3] Smetanina T, Pintassilgo P and Matias A 2014 *Economy and Law* **1** 79
[4] Degaltseva Yu, Atamas M and Lysenko E 2016 The XLIV International Scientific and Practical Conference “*Scientific community of students of the XXI century. Economic Sciences*”**7 (44)** 24
[5] Ormazabal M and Sarriegi J M 2012 *A Multiple Case. Study Industrial Engineering: Innovative Networks* (San Sebastian)
[6] Egorova O 2012 *Proceedings of the 23rd International DAAAM Symposium* **23 (1)** 1051

IOP Conf. Series: Materials Science and Engineering **843** (2020) 012005 doi:10.1088/1757-899X/843/1/012005

[7] M Fischer-Kowalski M and Swilling M 2011 *Decoupling Natural Resource Use and Environmental Impacts from Economic Growth* (UNEP International Resource Panel)
[8] Chuikova L and Chuikov S 2011 *Astrakhan Bulletin of Environmental Education* **3 (21)** 155
[9] Dvinin D 2014 *Management issues* **1 (7)** 194
[10] Osipov D 2011 *Databases and Delphi. Theory and practice* (SPb.: BHV-Petersburg)

IOP Conf. Series: Materials Science and Engineering **843** (2020) 012006 doi:10.1088/1757-899X/843/1/012006

Synthesis of irregular motion mechanisms for production machine drives

T G Martynova[1], V Yu Skeeba[1], Yu I Podgornyj[1] and D V Lobanov[2]

[1] Novosibirsk State Technical University, 20, K. Marx Str., Novosibirsk, 630073, Russia
[2] Ulianov Chuvash State University, 15, Moscow Ave., Cheboksary, 128010, Russia

E-mail: skeeba_vadim@mail.ru

Abstract. Irregular motion of the working bodies is widely used in mixing equipment. It helps to achieve homogeneity of the mixtures, since the uniform motion may lead to the "dead" zones formation. The purpose of this work is to synthesize the mechanism for driving the mixer's working shafts which includes an epicyclic transmission with a swing. This paper reports the studied pathways, velocities, and accelerations for different gear ratios at the final link, i.e. the swing. The research methods are analysis and synthesis of irregular motion mechanisms. It was found that in the mechanisms with the lowest pulsation, the amplitude of the angular acceleration change at the distance from the planet gear rotation axis to the pin equal to 14 mm is the smallest of the considered ones. The performance increases by 22 % with this coefficient of non-uniformity. As a result of the conducted research, we determined the main rational parameters for the synthesis of the considered mechanism.

1. Introduction

In many industries, mechanisms that impart irregular motion to working bodies are widely used. Such mechanisms are especially important in the mixing equipment for obtaining mixtures with uniform consistency.

The use of uniform motion in the mixers working bodies leads to the formation of the so-called "dead" zones in the working chamber, and, as a result, low quality of the finished product [1].

When an irregular motion is imparted to the working body, the actual instantaneous speed of the mixed raw material is composed of the average value of the flow rate over time and the deviation of the actual speed from the average (pulsation). This mode of particle movement allows achieving uniform consistency of the product [2 - 4].

This problem can be solved by the use of various drive mechanisms providing for irregular motion of the mixer working body [5 - 12].

In this paper, we consider the synthesis of the drive mechanism of the mixer's working shafts, which includes an epicyclic transmission with a swing.

2. Materials and methods

As a rule, mixers operate at a constant angular velocity of the working body. Under these conditions, over time the fluid velocity becomes equal to the speed of the blades of the mixer working body, i.e. there is a stratified fluid flow: particles of larger mass components move in orbits of a larger radius;

particles of smaller mass move in orbits of a smaller radius, thus creating laminar motion which negatively affects the results of the mixing process.

When a variable angular velocity is imparted to the working body, the actual instantaneous velocity of the fluid ω_f will be composed of the average value of the flow velocity over time ω_a and the deviation of the actual velocity from the averaged $\Delta\omega$:

$$\omega_f = \omega_a + \Delta\omega,$$

In this case, pulsations occur due to the deviation of the actual speed from its average value, which will differ in magnitude and direction. Consequently, turbulent phenomena occur, i.e. elementary liquid masses transition from one layer to another, which contributes to improving the quality and intensity of the mixing process.

The turbulent mode of particle motion allows achieving homogeneity of the mixture. Thus, to improve the quality of the mixture, it is advisable to use mixers with a variable angular rotation speed of the kneading body. This problem can be solved by using lever and gear drive mechanisms that ensure irregular motion of the mixer working body.

For these purposes, we propose to consider epicyclic transmission.

To synthesize such a mechanism, the geometric parameters of the constituent elements must be determined. The scheme of the mechanism is shown in Figure 1.

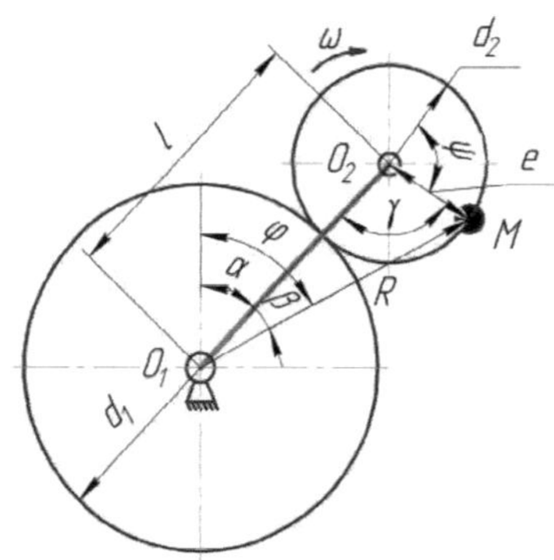

Figure 1. Epicyclic transmission: 1 – Sun gear; 2 – Planet gear; M – Pin tracing an epicycloid; d_1 – Diameter of the sun gear; d_2 – Diameter of the planet gear; l – Interaxial distance; e – Distance between the planet gear rotation axis and the pin M.

According to [2, 4, 6 - 12], the gear ratio is determined by the formula:

$$u = \frac{d_2}{d_1}, \tag{1}$$

The planet gear rotation angle

$$\psi = \frac{\alpha}{u}, \tag{2}$$

with α – the carrier rotation angle O_1O_2; u – gear ratio.

The angle $\gamma = \angle\, O_1O_2M$ is determined by:

$$\gamma = 180 \cdot \text{deg} - \psi, \tag{3}$$

The value of the radius vector R connecting points O_1 and M:

$$R = \sqrt{l^2 + e^e - 2 \cdot l \cdot e \cdot \cos\gamma}\,, \tag{4}$$

The angle $\beta = \angle O_2O_1M$ is found by the sine theorem:

$$\beta = \arcsin\left(\frac{e \cdot \sin(\psi)}{R}\right), \tag{5}$$

The angle φ is made up of the angle α and β:

$$\varphi = \alpha + \beta \tag{6}$$

The angular velocity of the epicyclic mechanism pin can be found by differentiating expression 6:

$$\omega_2 = \frac{d}{di}\varphi \tag{7}$$

The angular acceleration of the epicyclic mechanism pin is determined by differentiating the angular velocity equation (7):

$$\varepsilon_2 = \frac{d}{di}\omega = \frac{d^2}{di^2}\varphi \tag{8}$$

Pulsations are received from the equation:

$$\lambda_2 = \frac{d}{di}\varepsilon = \frac{d^2}{di^2}\omega = \frac{d^3}{di^3}\varphi \tag{9}$$

In order to impart the motion to the slave link, it is necessary to transfer the motion from the point M to the swing, whose axis coincides with the axis of the carrier rotation. Thus, the law of change in angular velocity and acceleration is transmitted to the working shaft.

Based on this, the pathways of the slave link with the transfer ratios u = 1; u = 1.5; u = 2; u = 3; u = 4 were obtained. The most characteristic pathways of pin motion (point M, see Figure 1) are shown in Figure 2.

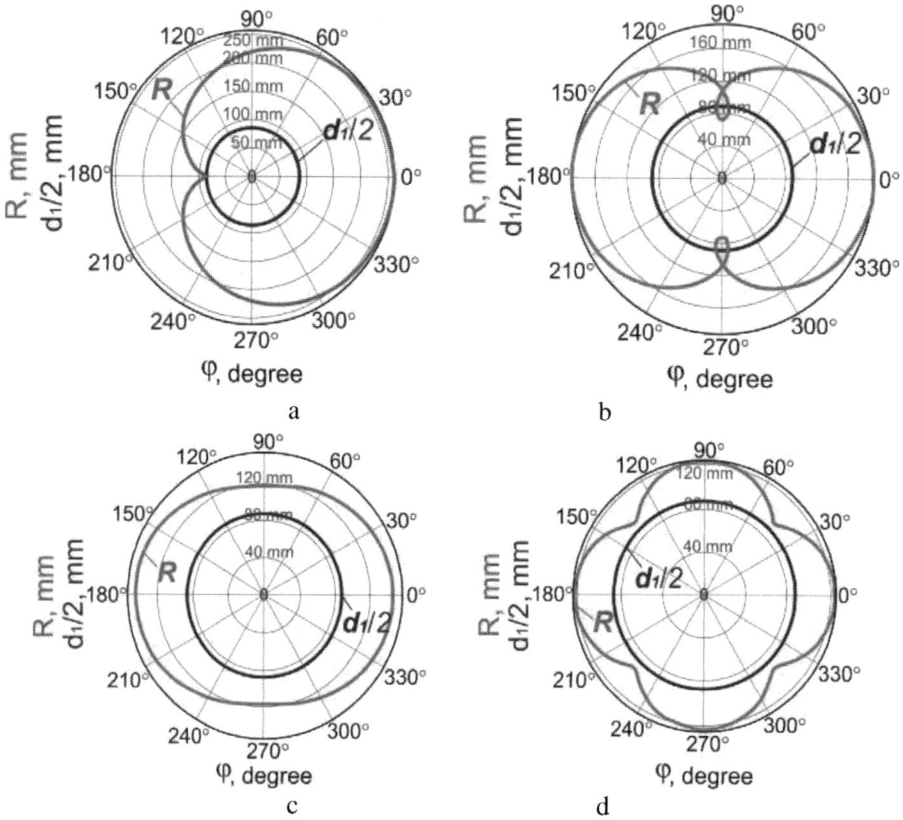

Figure 2. Pin motion pathways (M): a - u = 1; b - u = 1.5; c - u = 2; d - u = 4

Angular velocities and accelerations are shown in Figure 3.

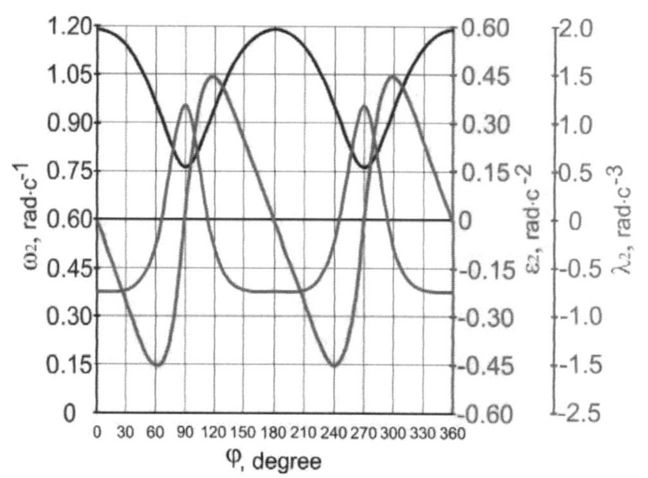

Figure 3. Diagram of angular velocities, accelerations, and pulsations: the transfer ratio u = 2.

3. Results and discussion

We demonstrate that the most rational gear ratio is $u = 2$ based on the analysis of graphs representing pathways, velocities, and accelerations at the final link (swing) at different gear ratios with the lowest angular accelerations.

Then the coefficient of non-uniformity δ is determined by the formula [1]:

$$\delta = \frac{\omega_{\max} - \omega_{\min}}{\omega_{ay}} , \tag{10}$$

with ω_{max}, ω_{min}, ω_{av} being the maximum, minimum, and average angular speeds of the swing, respectively.

The average angular velocity is calculated according to the expression:

$$\omega_{ay} = \frac{\omega_{\max} + \omega_{\min}}{2} , \tag{11}$$

The capability of a continuous paddle mixer C ($kg \cdot h^{-1}$) is determined by the formula [13]:

$$C = 1435 \cdot \varphi \cdot \rho \cdot \left(\frac{b}{H} \right) \cdot D^3 \omega \cdot \sum_{i=1}^{n} j_i \cdot \sin 2\alpha_i , \tag{12}$$

with $\varphi = 0.2$ – feed coefficient [13]; $\rho = 1200$ kg/m^3 – dough density [13]; $b = 48$ mm – blade width; $H = 300$ mm – blade pitch; $D = 0.5$ м – blade outer diameter; ω – angular speed of the working shaft; j_i – the number of blades mounted at one angle; α_i – the angle between the blades and the shaft center line; i – variants of the blades location (Figure 4).

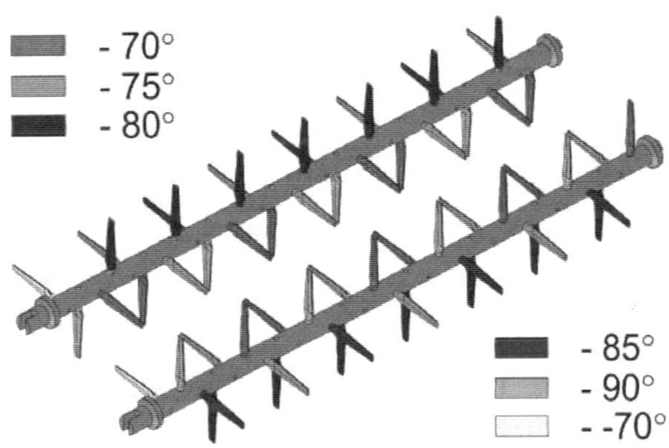

Figure 4. Blade mounting angles

Calculations for epicyclic transmission were made in accordance with formulas (1 - 11). On the basis of these values, we obtained the dependence of the machine performance on the nonuniformity coefficient (Figure 5).

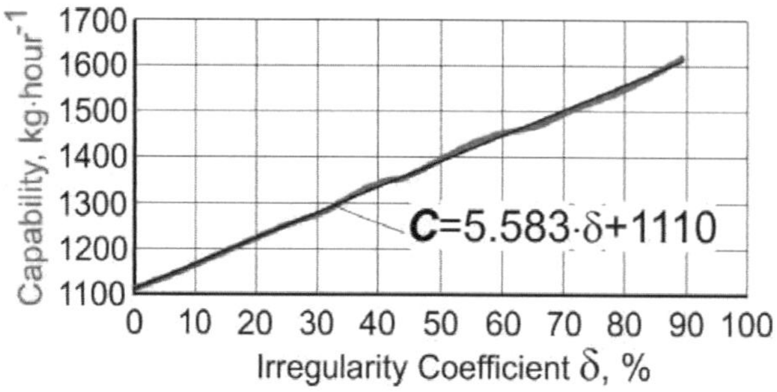

Figure 5. Dependence of the machine capability on the coefficient of rotation non-uniformity of the working body

It is obvious that there is a positive correlation between performance and the non-uniformity coefficient. Based on this, we determined the law of performance change related to the coefficient of non-uniformity. For this purpose, we build a trend line (Figure 5). The graph shows that when the non-uniformity coefficient δ increases by 1 %, the machine's performance increases by an average of 5.583 kg/h.

Based on the range of the recommended values of the non-uniformity coefficient [14], we choose the optimal values: $\delta_1 = 43.907$ %, $\delta_2 = 50.749$ %, $\delta_3 = 57.848$ % . These values correspond to the O_2M value (Figure 1): e_1 = 14 mm, e_2 = 16 mm, e_3 = 18 mm, respectively. Point M (Figure 1) traces an epicycloid.

The swing angular velocity is determined by the formula (7); the angular velocity of the carrier ω_1 equals to 1 rad/s (Figure 6).

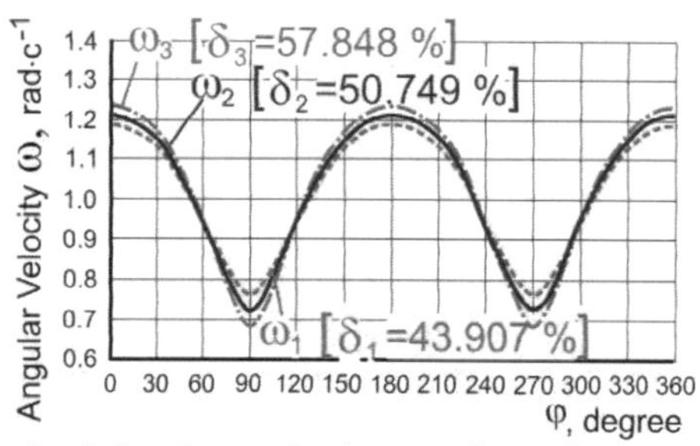

Figure 6. Graph of the change in the angular velocity of the swing

Angular acceleration is obtained by differentiating the angular velocity (formula 7); the graph of angular acceleration is shown in Figure 7.

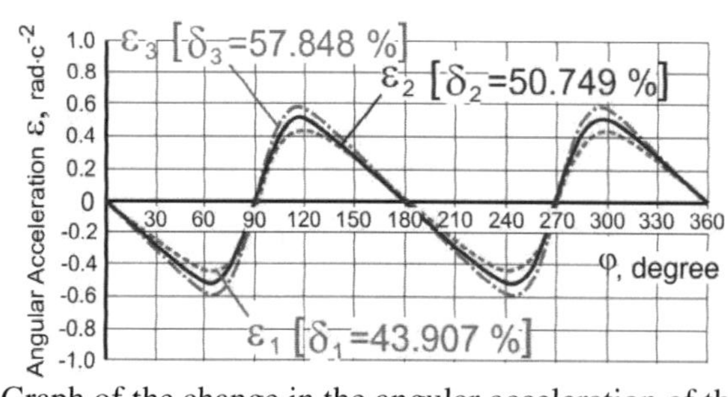

Figure 7. Graph of the change in the angular acceleration of the swing

The pulsation graph (Figure 8) is obtained by differentiating the angular acceleration (formula 9).

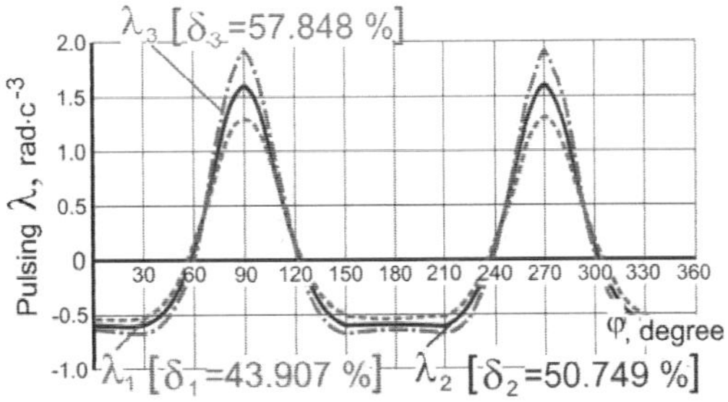

Figure 8. Graph of pulsations

4. Conclusion

In the course of the conducted research we investigated the epicyclic mechanisms with transfer ratios from 1 to 4 with a step of 0.5.

The mechanism with the lowest pulsation δ = 43.907% was chosen from the presented variants, since in this case the amplitude of the angular acceleration change at e = 14 mm is the smallest of the considered ones.

The performance increases by 22 % with this coefficient of non-uniformity.

Thus, the most rational basic parameters for the synthesis of an epicyclic mechanism with a swing are: the diameter of the sun gear d_1 = 176 mm; the diameter of the planet gear d_2 = 88 mm; the gear ratio u = 2; the interaxial distance l = 132 mm; the distance from the planet gear rotation axis to the pin e = 14 mm.

Acknowledgments

This work was financially supported within the framework of the NSTU Research and Development Thematic Plan (project No. TP-PTM-1_20 project).

References

[1] Mudrov A G 1999 *Development of spatial mixing devices of a new generation used in agriculture and industry* (Al'nopress Publ.)
[2] Hsieh J F 2014 *Mechanism and Machine Theory* **81** 155-165
[3] Ptitsyn S V et al. 2013 *Obrabotka metallov* **2** 33–38
[4] Myszka D H 2012 *Machines & Mechanisms: Applied Kinematic Analysis* (Pearson Publ.)
[5] Podgornyj Y I, Skeeba V Y, Kirillov A V et al. 2016 *Obrabotka metallov* **4** 24-33
[6] Zhoua C, Hua B, Chenb S, Mac L 2016 *Mechanism and Machine Theory* **104** 118-129
[7] Vulfson I 2015 *Dynamics of Cyclic Machines* (Springer Int. Publ.)
[8] Podgornyj Y I et al. 2019 *IOP Conf. Ser.: Earth Environ. Sci.* **378** **(1)** 012025
[9] Mott R L 2013 *Machine Elements in Mechanical Design* (Pearson Publ.)
[10] Podgornyj Y I et al. 2017 *IOP Conf. Ser.: Earth Environ. Sci.* **87** 082039
[11] Martynova T G, Skeeba V Y, Skeeba P Y 2018 *IOP Conf. Ser.: Earth Environ. Sci.* **194** **(2)** 022022
[12] Podgornyj Y I, Kirillov A V, Ivancivsky V V, Lobanov D V, Maksimchuk O V 2019 *Obrabotka metallov* **21** **(4)** 47-58
[13] Antipov S T 2001 *Machines and equipment for food production* (High School Press.)
[14] Matyashin A V 1998 *Justification of the parameters and operating modes of the feed mixer of periodic action* (KGSKhA Publ.)

IOP Conf. Series: Materials Science and Engineering **843** (2020) 012007 doi:10.1088/1757-899X/843/1/012007

Effect of aggressive media on working capacity of composite elements in mechanical engineering

S I Koryagin, O V Sharkov and N L Velikanov

Immanuel Kant Baltic Federal University, Kaliningrad, 236016, Russia

E-mail: [a]skoryagin@kantiana.ru

Abstract Composite elements can improve not only the strength but also the protective properties of engineering products, thus reducing corrosion losses. At present, there is no sufficient study of the environment influence on the working capacity of composite elements, in particular on the decrease of their crack resistance. The article discusses the calculation and experimental methodology, which allows increasing the accuracy and reliability of the assessment of crack resistance of composite elements under the influence of aggressive media. The presented methodology can be implemented without a test facility by testing directly experimental specimens. To determine the magnitude of the crack propagation energy, a mathematical model is proposed that includes only one independent parameter (radius of curvature). The authors investigated qualitative and quantitative characteristics of the effect on the fracture toughness of such aggressive media as machine oil and water. It is shown that the effect of air and machine oil on the crack resistance is approximately the same. A decrease in crack resistance by 1.56-11.1 times is observed under the influence of water. The composite elements based on the VAK-A adhesive composition have the highest crack resistance in aggressive media.

1. Introduction

The integrated effect of mechanical loads, temperature and aggressive media is a characteristic feature of the operation of various elements of metal engineering structures.

The elements of metal structures, used in various fields of mechanical engineering, undergo corrosion during operation. This results in a decrease in their bearing capacity and often premature decommissioning of facilities leading to significant losses of labor and material resources. For example, statistics from the American Petroleum Institute show annual corrosion costs of the oil and gas industry estimated at $ 2 billion. At the same time, the cost of corrosion control is about $ 500 million annually [1–4].

One of the promising solutions to this problem is the use of composite materials such as metal-reinforced polymer coating, in other words, the metal-polymer materials that combine the protective properties of coatings with the mechanical strength of the metal (substrate) [5–13].

However, the capabilities of such coatings have not yet been enough studied. In particular, there are no sufficiently proven recommendations on the use of polymer coatings under the influence of aggressive media (machine oil, water and etc.).

One of the main working capacities of the polymer coatings (bearing capacity, durability, etc.) is crack resistance, which depends on the amount of adhesion – the bonding force of the adhesive with the matrix.

Currently, a large number of theoretical and experimental methods is used to determine the crack resistance of polymer coatings and adhesive joints [14–24].

A major drawback of existing experimental techniques is the need for test facilities. Thus, they cannot be directly used to assess crack resistance under the influence of aggressive media on polymer coatings.

A modified calculation and experimental methodology has been developed for assessing the crack resistance of polymer coatings operating in aggressive media without the use of a testing facility.

2. Methodology for assessing crack resistance of composite elements

2.1. Methodology description

The essence of the proposed method lies in the fact that the experimental specimen consists of two beams made in the form of a circular arc. They are forcibly straightened and connected using an adhesive joint. Then the straightener is removed. The beams tend to take their original shape which results in a crack formation in the adhesive line (Figure 1).

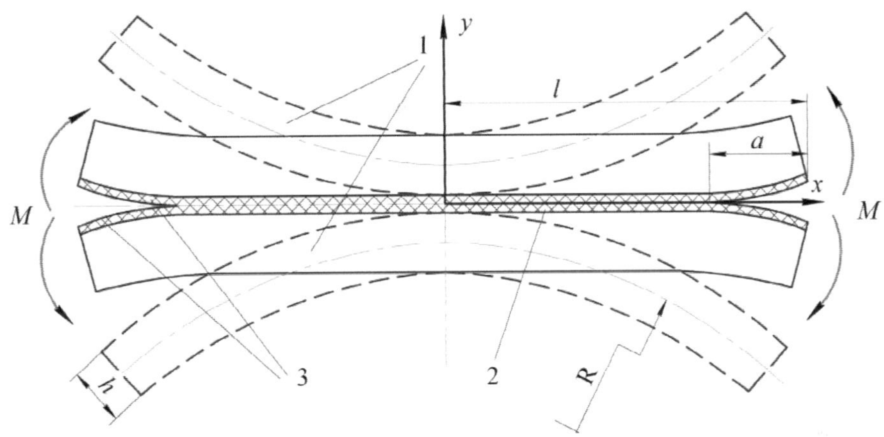

Figure 1. Crack formation pattern: 1 – beams (the shape of beams before connection is shown by dashed lines); 2 – adhesive composition; 3 – crack

The implementation of the developed methodology is as follows. Beams are preliminarily bent so that the displacements are in the zone of elastic deformations. For example, a three-roll bending machine can be used.

After this, the beams acquire a curvilinear shape and their radius of curvature is described by an arc of a circle with a radius R.

To eliminate the residual stresses arising in the beams due to the bending of the specimen, the heat treatment is performed.

Then the beams are straightened using special brackets, and an adhesive composition is applied to their inner surface. Then the beams are joined together and held for a certain time until the composition is cured.

Upon that, the brackets are removed and the beams tend to acquire their original shape. Due to the action of the elastic moment M, the experimental specimen is wedged from two sides, as a result cracks are formed.

Thus, the pre-bent beams allow the loading of the polymer coating without the use of special devices.

It is recommended to pre-bend the beams in a way that the neutral axis of their individual sections had the shape of arcs of circles with different radii. In this case, the constant magnitude condition of the cracking force within the sample length section is achieved.

2.2. Mathematical models and calculation algorithm

To assess crack resistance, the crack propagation rate $\dot{a}$ is taken as a dependent parameter, and the crack propagation energy G is taken as an independent parameter [19–22].

The initial data for determining the crack propagation energy is a system of expressions that includes the dependences for determining the potential energy, bending moment, and radius of curvature [25, 26]:

$$\begin{cases} U = \int\limits_{l_1}^{l_2} \frac{M^2(x)}{2EJ} dx; \\ M(x) = \frac{d^2 y}{dx^2} EJ; \\ R \approx \frac{1}{d^2 y / dx^2}, \end{cases} \quad (1)$$

where E – module of elasticity of beams; J – moment of inertia of the cross sections of cantilever beam.

The joint solution of the system of expressions (1) allows obtaining the dependence for determining the potential energy of the adhered specimen

$$U = \int\limits_{a}^{l} \left[\frac{1}{2R^2(x)} \left(E_1 J_1 + k^2 E_2 J_2 - \frac{(E_1 J_1 - k E_2 J_2)^2}{(EJ)_0} \right) \right] dx, \quad (2)$$

where R – the radius of curvature of the beams before adhering; k – coefficient depending on the characteristics of the material and the size of beams; $(EJ)_0$ – bending stiffness of a unit length of the glued beams.

The expression for determining the energy of crack propagation can be written as

$$G = -\frac{1}{b} \frac{dU}{da}. \quad (3)$$

where b – the width of the adhesive seam.

Taking into account the dependences (2) and (3) we obtain the expression for the crack propagation force. We assume that the experimental specimen consists of two cantilever beams made with the same moduli of elasticity ($E_1 = E_2 = E$) and inertia ($J_1 = J_2 = J$).

Then the expression for determining the crack propagation energy with the accepted assumptions and dependencies (7) and (8) can be written as follows

$$G = \frac{1}{2bR^2} \left[EI(1+k) \frac{[(1-k)EI]^2}{(EI)_0} \right]. \quad (4)$$

The expression is convenient for use, since it contains only one independent parameter – the radius of curvature R, which can be determined through the deflection of the beam as

$$R = \frac{l^2}{8f},$$

where f – the deflection of the beam; l – the base length of the beam when measuring the magnitude of the deflection.

The measurement of crack opening a can be done using a microscope.

3. Results of calculation and experimental studies

The studies used polymer coatings based on the following adhesive compositions: VAK-A, Sprout-9M, Sprout-Plus, the characteristics of which are given in the Table 1. In the experimental specimen, St 3 steel was a substrate and T-II-GVS-9 fiberglass was a reinforcing material.

Table 1. Technological characteristics of adhesive compositions

Physical and chemical properties	Adhesive composition		
	Sprout-9M	Sprout-Plus	VAK-A
Strength in air at:			
uniform separation, MPa	7.5	8.4	10.8
shear, MPa	7.0	7.0	10.1
Strength in water at:			
uniform separation, MPa	4.6	5.7	7.7
shear, MPa	4.1	5.2	7.3
Strength in machine oil at:			
uniform separation, MPa	6.5	8.6	8.6
shear, MPa	5.4	7.8	8.2

The polymer coatings were applied in air. Then the experimental specimens were kept for 90 hours in various media – in air, in machine oil and water. After that, the kinetics of crack growth was evaluated.

Figures 2–4 present diagrams showing the kinetics of subcritical crack growth in polymer coatings under the influence of various media, obtained on the basis of experimental data processing.

The figures show the upper limit of the rate of crack propagation according to the averaged values of the experimental points.

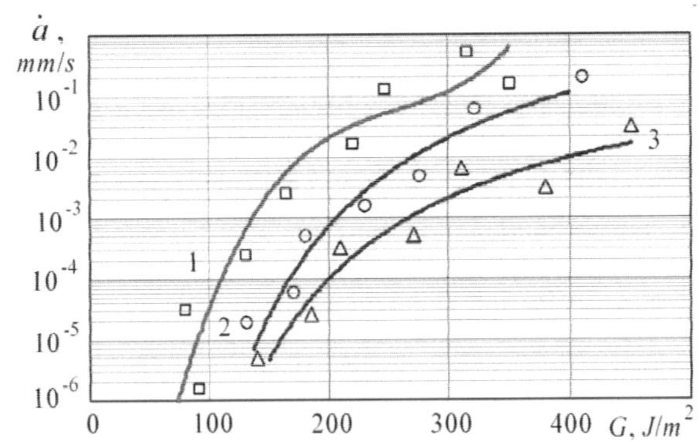

Figure 2. Pattern of the air effect on the destruction of various adhesive compositions: 1 – Sprout-9M; 2 – Sprout-Plus; 3 –VAK-A

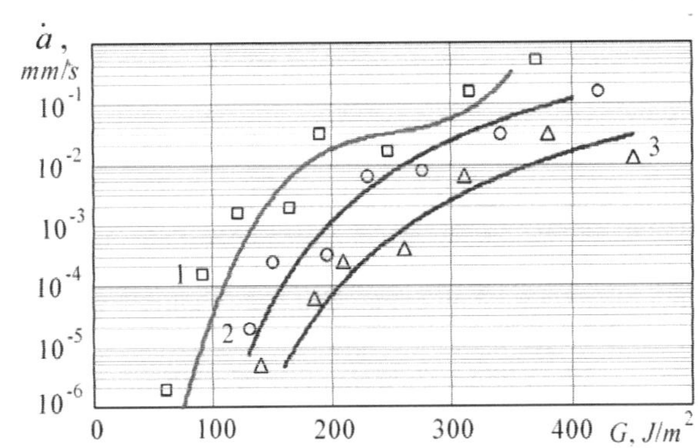

Figure 3. Pattern of the machine oil effect on the destruction of various adhesive compositions: 1 – Sprout-9M; 2 – Sprout-Plus; 3 –VAK-A

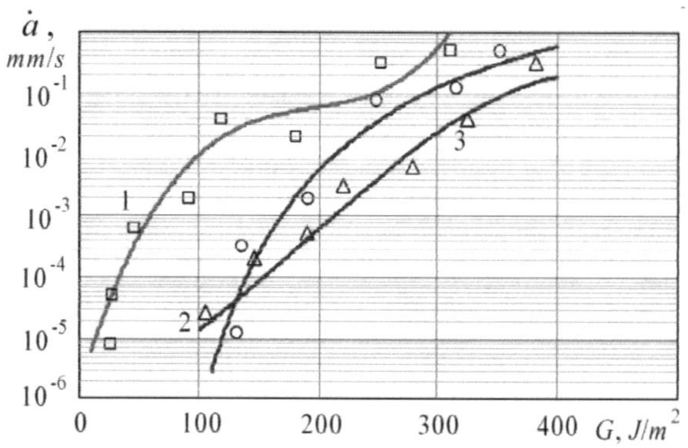

Figure 4. Pattern of water effect on the destruction of various adhesive compositions: 1 – Sprout-9M; 2 – Sprout-Plus; 3 –VAK-A

Table 2 compares the crack resistance characteristics of polymer coatings made on the basis of various adhesive compositions.

Table 2. Parameters of crack resistance of adhesive compositions

Energy of crack propagation, J/m^2	Average speed of crack propagation, mm/c		
	Sprout-9M	Sprout-Plus	VAK-A
		Water	
100	$1.0 \cdot 10^{-2}$	–	$1.1 \cdot 10^{-5}$
200	$1.8 \cdot 10^{-2}$	$1.5 \cdot 10^{-3}$	$1.7 \cdot 10^{-4}$
300	$1.5 \cdot 10^{-1}$	$1.1 \cdot 10^{-1}$	$1.2 \cdot 10^{-2}$
400	–	$1.6 \cdot 10^{-1}$	$1.2 \cdot 10^{-1}$
		Air	
100	$1.3 \cdot 10^{-5}$	–	–
200	$1.2 \cdot 10^{-2}$	$1.9 \cdot 10^{-4}$	$1.0 \cdot 10^{-4}$
300	$1.0 \cdot 10^{-1}$	$1.2 \cdot 10^{-2}$	$1.2 \cdot 10^{-3}$
400	–	$1.1 \cdot 10^{-1}$	$1.0 \cdot 10^{-2}$
		Machine oil	
100	$1.2 \cdot 10^{-5}$	–	–
200	$1.1 \cdot 10^{-2}$	$1.0 \cdot 10^{-3}$	$1.8 \cdot 10^{-5}$
300	$1.5 \cdot 10^{-2}$	$1.2 \cdot 10^{-2}$	$1.2 \cdot 10^{-3}$
400	–	$1.1 \cdot 10^{-1}$	$1.1 \cdot 10^{-1}$

The statistical processing of the experimental results showed: the relative error of a series of measurements a varies in the range of 9.3...17.5%; the test variance can be considered homogeneous, since the maximum values of the Cochren criterion for all series of measurements do not exceed critical $G = 0.390 < G_{0,95} = 0.480$ [27].

4. Conclusion

The test results showed that machine oil have a very small effect on the rate of crack propagation in comparison with the air.

This is explained by the fact that machine oil have a higher molecular weight, and this is an obstacle to their diffusion into the adhesive composition, as well as the lack of chemical interaction of the composition with machine oil.

Specimen tests in water showed that the effect of water on the adhesive joint leads to an increase in the rate of crack propagation by 1.56-11.1 times (in the range G =200…300 J/m^2), depending on the size G and type of adhesive composition.

Obviously, water causes chemical changes in the adhesive composition. This is the main reason for the decrease in crack resistance of polymer coatings and adhesive compounds under the influence of water. The greatest crack resistance when exposed to aggressive media has a composite element based on the VAK-A adhesive composition. The crack resistance of the Sprout-9M and Sprout-Plus adhesive compositions is lower.

References

[1] Rajaguru J and Arunachalam N 2018 *Corrosion Science* **14115** 230–242
[2] Prakash M, Shekhar S, Moon A P and Mondal K 2015 *Journal of Materials Processing Technology* **219** 70–83
[3] Zaki A 2006 *Principles of Corrosion Engineering and Corrosion Control* (Amsterdam: Elsevier)
[4] Kaesche H 2003 *Corrosion of Metals: Physicochemical Principles and Current Problems* (Berlin: Springer Science & Business Media)
[5] Vipin V, Sambhu R and Raphel D C 2019 *Materials Today: Proceedings* **18 (7)** 3749–3757
[6] Meneghetti G, Quaresimin M and Ricotta M 2012 *Composites Part B: Engineering* **43 (2)** 210–220
[7] Koryagin S I, Sharkov O V and Velikanov N L 2018 *Materials Science Forum* **938** 46–53
[8] Cho S-K, Kim H-J and Chang S-H 2011 *Composite Structures* **93** 492–501
[9] Dubrovskiy A, Aliukov S. Dubrovskiy S and Alyukov A 2018 *Engineering Letters* **26 (4)** 526–534
[10] Camanho P P and Lambert M 2006 *Composites Science and Technology* **66** 3004–3020
[11] Smolik J and Kulisek V 2009 *Journal of Machine Engineering* **9 (2)** 93–105
[12] Chai H and Fox J 2012 *International Journal of Solids and Structures* **49 (22)** 3142–3147.
[13] Badie M A, Mahdi E and Hamouda A M S 2011 *Materials&Design* **32 (3)** 1485–1500
[14] Kuhn P, Catalanotti G, Xavierc J, Camanho P P and Koerber H 2017 *Composite Structures* **182** 164–175
[15] Chai H and Fox J 2012 *International Journal of Solids and Structures* **49 (22)** 3142–3147
[16] Yang C-W and J-W Park 2010 *Surface and Coatings Technology* **204 (16–17)** 2761–2766
[17] Catalanotti G, Arteiro A, Hayati M and Camanho P P 2014 *Engineering Fracture Mechanics* **118** 49–65
[18] Koryagin S I, Velikanov N L and Sharkov O V 2017 *Key Engineering Materials* **736** 8–11
[19] Gift M D M, Selvakumar J and Alexis S J 2015 *Journal of Chemical and Pharmaceutical Sciences* **2015-April** 269–273.
[20] Zesers A and Tamuzs V 2014 *Mechanics of Composite Materials* **50 (2)** 165–176
[21] Alexeev A A, Bolshev K N, Ivanov V A, Syromyatnikova A S and Bolshakov A M 2019 *Procedia Structural Integrity* **20** 254–258
[22] Matvienko Yu G, Konovalov V V and Pankov A V 2015 *Journal of Machinery Manufacture and Reliability* **44 (6)** 520–530
[23] Gowthama E, Velmurugan V, Paramasivam V and Thanikaikarasan S 2020 *Materials Today: Proceedings* **21(1)** 694–700
[24] Yang Y, Ju Y, Li F, Gao F and Sun H 2016 *Journal of Natural Gas Science and Engineering* **32** 415–422
[25] Case J and Chilver A H 2013 *Strength of Materials and Structures: An Introduction to the Mechanics of Solids and Structures* (Amsterdam: Elsevier)
[26] Gambhir M L 2011 *Fundamentals of Structural Mechanics and Analysis* (New Delhi: PHI Learning Pvt. Ltd)
[27] Mandel J 2012 *The Statistical Analysis of Experimental Data* (New York: Dover Publications)

Stress state of two-layer composite elements of curved shape

S I Koryagin, O V Sharkov and N L Velikanov

Immanuel Kant Baltic Federal University, Kaliningrad, 236016, Russia

E-mail: skoryagin@kantiana.ru

Abstract. Two-layer composite structures consisting of a metal base and a polymer coating are widely used in mechanical engineering. To assess the bearing capacity and durability of such structures, methods for calculating their stress-strain state are necessary. The paper theoretically investigates the stress state of a two-layer structural element of a curved profile for two variants - under pure bending and under temperature change. The radial and meridional stresses arising in the polymer coating and the metal base of the element are taken as the studied factors. The geometric, strength, and physical characteristics of the structural element are accepted as the independent factors. Mathematical models are obtained using the plane section hypothesis. The authors investigated how the radial and meridional stresses acting in the polymer coating are related to the ratio of the coating and the metal thicknesses, as well as the metal thickness to the curvature radius. It has been established that the curvature of the metal element and the ratio of the layers thickness have the most significant effect on the magnitude of the radial stresses, and the ratio of the thicknesses on the meridional stresses. The stress state of the polymer coating is mainly determined by the ratio of the metal thickness to the radius of curvature of the boundary layer.

1. Introduction

Nowadays composite materials based on polymer coatings are widely used in the manufacture of various structural elements in mechanical engineering, aircraft manufacturing and other industries. Composite materials are efficient due to their lightness, variability of composition and structure, and, consequently, a wide range of technical properties [1–6].

A characteristic property of a significant number of structural elements, especially in the aircraft industry, is that they are formed by curved surfaces. Examples of the composite materials used by aircraft manufacturers in the USA and Great Britain include: brake flaps for slats, stabilizers (F-5 and Skyhawk A-4 aircraft); a wing of BQM-34F target drone; retraction system doors (CF-14 reconnaissance aircraft), transmission shafts (Wasp/Scout helicopter); external paneling, spars, attachment points and landing gear; monocoque tail booms and other parts for other aircraft and helicopters. According to preliminary estimates, the widespread use of composite materials can reduce the mass of: helicopters by 35%; military transport aircraft by 22%; vertical take-off and landing aircraft by 21% [7–10].

This, in turn, has a significant effect on the stress-strain state of polymer coatings applied to a curved metal base. It should also be borne in mind that during operation, metal structures can be subjected to intense shock, vibration, and temperature loads.

The literature has limited information on stresses in a polymer coating on a curved metal surface. Existing calculation methods are applicable when the metal-coating composition is deformed without bending during the formation [11-20]. But in real structural elements, areas with a large curvature are, as a rule, in conditions of moment stress state.

This makes theoretical and experimental research relevant in order to assess the stress state of the reinforced polymer coating on the curved surface of metallic mechanical engineering structural elements.

2. Methodology for assessing the stress state of two-layer curved composite elements

2.1. Calculation scheme and mathematical models

Consider the areas located far enough from the coating edge, which allows neglecting the effect of stress concentration at the edge. The selected curved composite element of unit width is bounded by two meridional sections and two sections normal to the meridian, the angle between which is $d\theta$. The composite element is a two-layer curved beam (Figure 1), which curvature can be considered constant in the area corresponding to the angle $d\theta$.

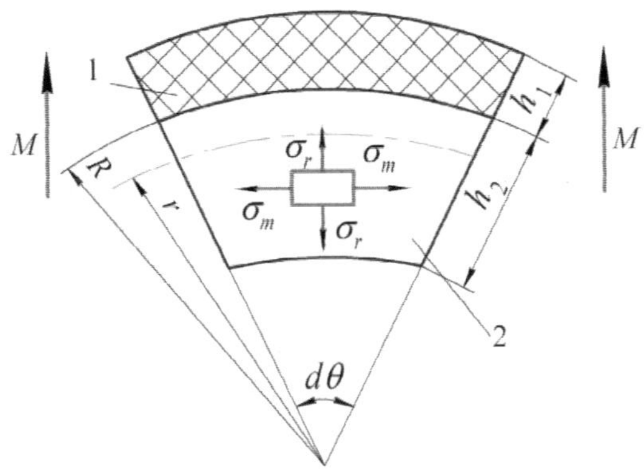

Figure 1. Two-layer composite curved composite element: 1 – polymer coating; 2 – metal construction.

Let us consider two cases of stresses arising in a two-layer beam: under pure bending and under temperature change. These cases correspond to the main types of stresses in the coating – operational and technological (residual). In solving this problem, we accept the following hypothesis: a flat section remains flat even after deformation of a two-layer beam, but it can stretch or contract in the radial direction. This assumption makes it possible to take into account radial stresses σ_r.

The accepted hypothesis establishes dependence [21]

$$\frac{d}{d_r}(\varepsilon_m r)-\varepsilon_r=\frac{\Delta d\theta}{d\theta}, \tag{1}$$

where r – the radius of curvature of the beam`s curved axis; ε_r – the relative strain acting in the radial direction (radial relative strain); ε_m – the relative strain acting parallel to the metal-coating interface (meridional relative strain); $\Delta d\theta$ – the change in angle as a result of the beam strain.

Using Hooke's law [21, 22] for the biaxial stress state and the equilibrium equation, we write

$$\sigma_m=\frac{d}{d_r}(\varepsilon_r r) \tag{2}$$

and taking into account the expression (1), we obtain:

$$r^2\frac{d^2\sigma_r}{dr^2}+3r\frac{d\sigma_r}{dr}=E\frac{\Delta d\theta}{d\theta}, \tag{3}$$

where σ_m – the meridional stress, E – the modulus of elasticity of the two-layer curved beam material.

We denote the polymer layer of the curved beam by index 1, and the metal layer by index 2 (Fig. 1). Then the solution of equation (3) for radial stresses can be written as:

$$\sigma_{ri} = 0.5E_i \frac{\Delta d\theta}{d\theta} \ln r + a_i r^{-2} + b_i \ , \qquad (4)$$

where i = 1, 2.

From the equilibrium equation we obtain the equation for determining the meridional stresses:

$$\sigma_{ri} = 0.5E_i \frac{\Delta d\theta}{d\theta}(1 + \ln r) + a_i r^{-2} + b_i \, . \qquad (5)$$

To find a_i, b_i and $\Delta d\theta / d\theta$ constants, we use the boundary conditions and the compatibility condition for the strain of the polymer and metal layers on the their interface. The boundary conditions lies in the radial stresses σ_r=0 on the curved surfaces of the two-layer beam and the stresses are reduced to the M moment on the end surfaces.

2.2. *Assessment of the stress state under pure bending*

We obtain expressions for determining radial and meridional stresses in the two-layer curved structural element under pure bending.

We introduce additional notation: $k_1 = 1 + (h_1/R)$, $k_2 = 1 - (h_2/R)$, $k_3 = 1 + (h_2/R)$, $k_4 = r/R$, $k_5 = 1 - \mu_1$, $k_6 = 1 - \mu_2$, $k_7 = 1 - (h_1/R)$, $k_8 = 1 + \mu_2$, $k_9 = 1 + \mu_1$, $e = E_1/E_2$,

where h_1 and h_2 – the coating and metal thickness; R – the radius of curvature of the boundary layer; μ_1 and μ_2 – the Poisson`s ratios for the coating and metal, respectively; E_1 and E_2 – the tensile moduli of the coating and metal material.

After determining the constants and solving equations (4) and (5), these expressions are written in the form:

$$\sigma_{r1} = \frac{-Me^{(h_1/R)}}{Z_1 h_1^2}\left[D_1\left(\frac{1}{k_4^2} - \frac{1}{k_1^2} \right) + \ln \frac{k_4}{k_1} \right]; \qquad (6)$$

$$\sigma_{m1} = \frac{-Me^{(h_1/R)}}{Z_1 h_1^2}\left[-D_1\left(\frac{1}{k_4^2} + \frac{1}{k_1^2} \right) + \ln \frac{k_4}{k_1} + 1 \right]; \qquad (7)$$

$$\sigma_{r2} = \frac{-Me^{(h_2/R)}}{Z_1 h_2^2}\left[D_2\left(\frac{1}{k_4^2} - \frac{1}{k_2^2} \right) + \frac{1}{e} \ln \frac{k_4}{k_2} \right]; \qquad (8)$$

$$\sigma_{m2} = \frac{-Me^{(h_2/R)}}{Z_1 h_2^2}\left[-D_2\left(\frac{1}{k_4^2} + \frac{1}{k_2^2} \right) + \frac{1}{e}\left(1 + \ln \frac{k_4}{k_2} \right) \right]. \qquad (9)$$

Here: $Z_1 = e[D_1 \ln k_1 + D_2 \ln k_2 + 0.25(k_1^2 - k_2^2)]$;

$$D_1 = \frac{(e \ln k_1 - \ln k_2)(1 + \mu_2 + k_2^{-2} k_6) + (k_5 \ln k_1 - k_6 \ln k_2)(1 - k_2^{-2})}{e(1 - k_1^{-2})(1 + \mu_2 + k_2^{-2} k_6) - (1 - k_2^{-2})(1 + \mu_2 + k_2^{-2} k_6)};$$

$$D_2 = \frac{(1 - k_2^2)(k_5 \ln k_7 - k_6 \ln k_2) + (\ln k_1 - e^{-1} \ln k_2)(1 + \mu_2 + k_3^{-2} k_5)}{e(1 - k_1^{-2})(k_8 + k_2^{-2} k_6) - (1 - k_2^{-2})(k_8 + k_7^{-2} k_5)}.$$

The direction of moment M indicated in Figure 1 is taken as positive value. Indices 1 and 2 in expressions (6)–(9) relate to the coating and metal layers, respectively.

2.3. Assessment of stress state under temperature change

When the temperature of the two-layer curved beam changes by variable Δt due to differences in the mechanical properties of the layer material, technological stresses arise.

We introduce notation $\eta_1 = (\alpha_1 - \alpha_2)\Delta t E_1$ and $\eta_2 = (\alpha_1 - \alpha_2)\Delta t E_2$,

where α_1 and α_2 – the linear expansion coefficients for the coating and metal, respectively.

Then, dependencies (4) and (5) for determining radial and meridional stresses take the form:

$$\sigma_{r1} = \frac{\eta_1}{Z_2}\left[\frac{D_3}{e}\left(\frac{1}{k_4^2} - \frac{1}{k_1^2}\right) + \ln\frac{k_4}{k_1}\right]; \quad (10)$$

$$\sigma_{m1} = \frac{\eta_1}{Z_2}\left[-\frac{D_3}{e}\left(\frac{1}{k_4^2} + \frac{1}{k_1^2}\right) + \ln\frac{k_4}{k_1} + 1\right]; \quad (11)$$

$$\sigma_{r2} = \frac{\eta_2}{Z_2}\left[D_4\left(\frac{1}{k_4^2} - \frac{1}{k_2^2}\right) + \ln\frac{k_4}{k_2}\right]; \quad (12)$$

$$\sigma_{m2} = \frac{\eta_2}{Z_2}\left[-D_4\left(\frac{1}{k_4^2} + \frac{1}{k_2^2}\right) + \ln\frac{k_4}{k_2} + 1\right]. \quad (13)$$

Here: $Z_2 = \frac{D_3}{e}(k_9 + k_1^{-2}k_5) + D_4(k_8 + k_2^{-2}k_6)(k_6 \ln k_2 - k_5 \ln k_1)$;

$$D_3 = \frac{(1-k_2^{-1})[0.25(ek_1^2 - 1 + k_2^2)] - (e\ln k_1 - \ln k_2)\ln k_2}{(1-k_2^{-2})\ln k_1 - (1+k_1^{-2})\ln k_2};$$

$$D_4 = \frac{-(e\ln k_1 - \ln k_2)\ln k_1 + (0.25[e(k_1^2 - 1) + 1 - k_2^2)](1-k_1^{-2})}{(1-k_2^{-2})\ln k_1 - (1-k_1^{-2})\ln k_2}.$$

3. Results of calculation and discussion

Figures 2–7 show the results of calculations according to dependences (6)–(13), which describe radial stresses under pure bending and temperature changes for the coating applied to a convex surface of a metal element provided: $e = E_1 = 0{,}1E_2$; μ_1=0,15; μ_2=0,35.

The nature of the change in radial and medial stresses in a reinforced polymer coating depending on the ratio of the layer thickness of the polymer coating h_1 and the metal h_2 under temperature change is shown in Figure 2 and 3.

As it can be seen from the results obtained, radial stresses σ_{r1} depend on the curvature of the metal element and the ratio of the thicknesses of the layers, and meridional stresses σ_{m1} mainly depend on the ratio of thicknesses. The increase of σ_{r1} and decrease of σ_{m1} are, apparently, one of the reasons for the fact known in practice: the strength of the coating on the curved surface of the structural element decreases with increasing thickness.

The nature of the change in radial and medial stresses in a reinforced polymer coating depending on the ratio of the metal thickness h_2 to the radius of curvature of the boundary layer R under pure bending and temperature change is shown in Figure 6–7.

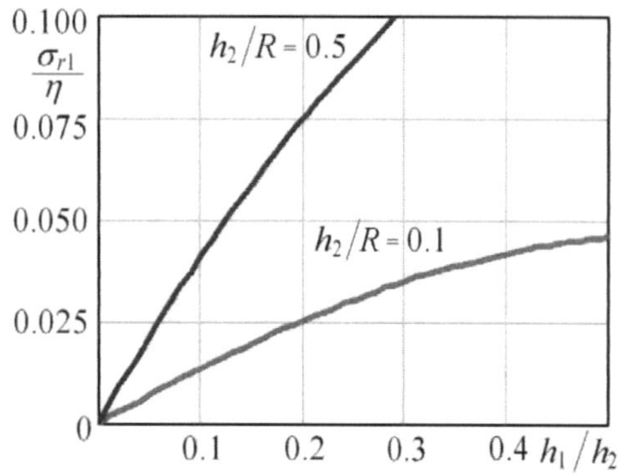

Figure 2. Dependence of radial stresses on the ratio of the layer thicknesses of the coating and the metal under temperature change.

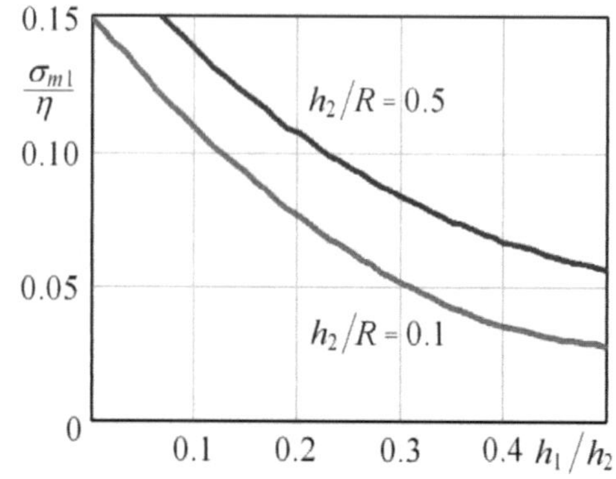

Figure 3. Dependence of meridional stresses on the ratio of the layer thicknesses of the coating and the metal under temperature change.

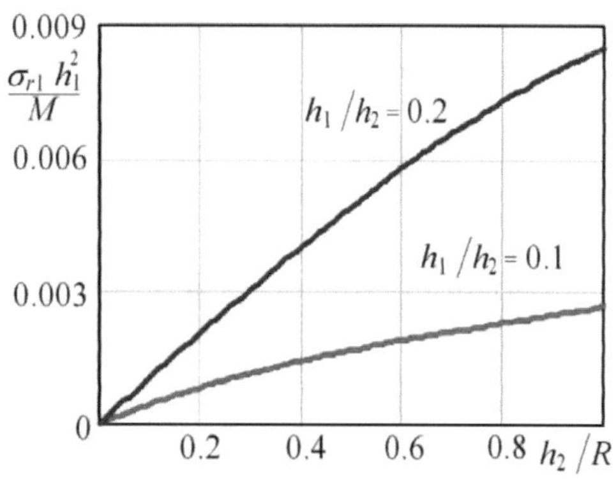

Figure 4. Dependence of radial stresses on the ratio of the thickness of the metal to the radius of curvature under pure bending.

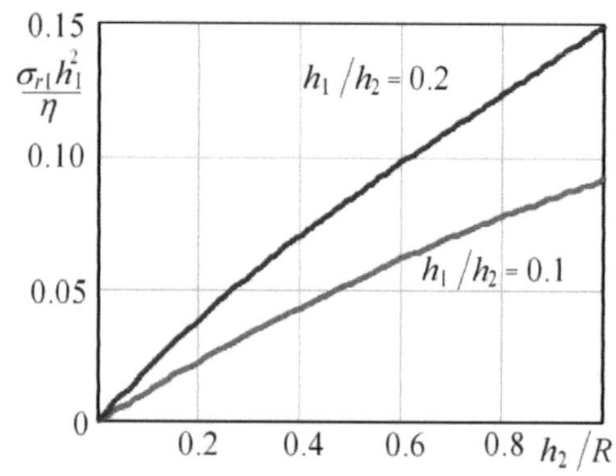

Figure 5. Dependence of radial stresses on the ratio of the thickness of the metal to the radius of curvature under temperature change.

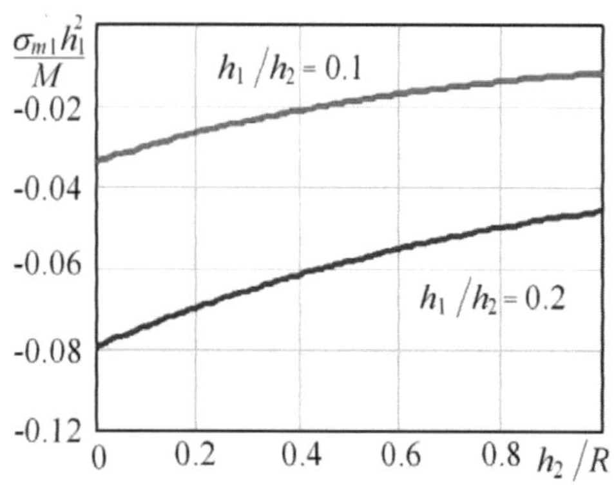

Figure 6. Dependence of meridional stresses on the ratio of the thickness of the metal to the radius of curvature under pure bending.

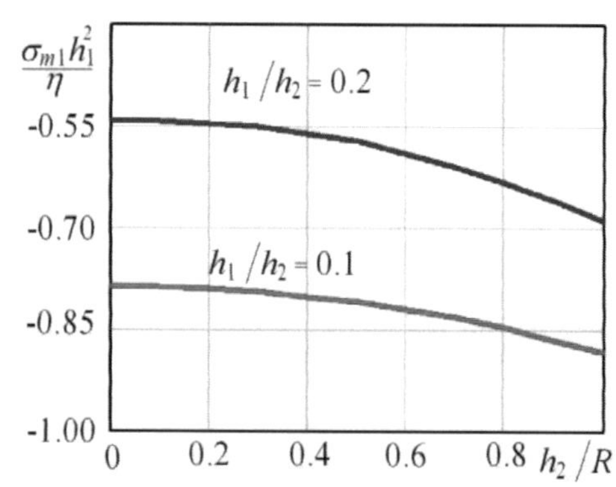

Figure 7. Dependence of meridional stresses on the ratio of the thickness of the metal to the radius of curvature under temperature change.

The analysis of the curves shows that tensile stresses σ_{r1} normal to the adhesive surface increase with increasing curvature. This poses a risk of normal cleavage of the coating from the metal surface.

At the same time, the meridional stresses σ_{m1} on the free surface of the coating are reduced, which weakens the resistance of the coating to stretching strain. This pattern of stress variation is observed both under pure bending and temperature change.

4. Conclusion

Assuming that the plane section hypothesis is fulfilled, a mathematical model is obtained – dependences (6)–(13), which describe the stress state of the polymer coating.

When determining radial and meridional stresses in a two-layer curved mechanical engineering component, we took into account pure bending and technological stresses arising in the materials of the layers of a two-layer curved beam under temperature change.

According to the obtained mathematical models, the ratio of the metal`s thickness to the radius of curvature of the boundary layer h_2/R is the most important parameter on which the stress distribution in the polymer coating depends.

Tensile stresses normal to the adhesive surface increase with increasing curvature, which can lead to cleavage of the coating from the metal surface. Stresses on the free surface of the coating are reduced, which weakens the resistance of the coating to stretching strain.

References

[1] Diaz O G, Luna G G, Liao Z and Axinte D 2019 *International Journal of Machine Tools and Manufacture* **139** 24–36
[2] Baumeister E, Klaeger S and Kaldos A 2004 *Journal of Materials Processing Technology* **155–156** 1839–1846
[3] Kulíšek V, Růžička M, Vrba P, Smolík J and Janota M 2019 *Materials Today: Proceedings* **12 (2)** 288–297
[4] Alyukov S V 2014 Russian Engineering Research **34 (9)** 549–553
[5] Möhring H-C 2017 *Procedia CIRP* **66** 2–9
[6] Irisarri F X, Macquart T, Julien C and Espinassou D 2019 *Composites Part B: Engineering* **158** 364–372
[7] Eken S 2019 *Thin-Walled Structures* **139** 362–371
[8] Dutton S, Kelly D and Baker A 2004 *Composite Materials for Aircraft Structures* (Reston, Virginia: AIAA)
[9] Marsh G 2016 *Reinforced Plastics* **60 (5)** 302–305
[10] Molent L and Forrester C 2017 *Composite Structures* **166** 22–26
[11] Yuan M, Wang Z, Yao Y and Li L 2019 *Results in Physics* **15** 102706
[12] Duan S, Zhang Z, Wei K, Wang F and Han X 2020 *Composite Structures* **236** 111884
[13] Ojo S O, Patni M and Weaver P M 2019 *International Journal of Solids and Structures* **178–179** 145–166
[14] Niu Y, Li X, Zhitao M and Liu T 2016 *Journal of Adhesion Science and Technology* **30 (22)** 2485–2497
[15] Koryagin S I, Sharkov O V and Velikanov N L 2018 *Materials Science Forum* **938** 46–53
[16] Shiah Y C and Hsu K W 2019 *Composites Part B: Engineering* **159** 105–113
[17] Fernández R, Cabeza S, Mishurova T, Fernández-Castrillo P and Bruno G 2018 *Materials Science and Engineering: A* **731** 344–350
[18] Strizhius V E 2019 *Mechanics of Composite Materials* **55 (4)** 495–502
[19] Marchuk A V and Nishchota A V 2018 *International Applied Mechanics* **54 (3)** 315–330
[20] Groha R M J and Tesslerb A 2017 *Computer Methods in Applied Mechanics and Engineering* **320** 369–395
[21] Pattabiraman J 2019 *Strength of Materials* (Chinnai-Tiruwelveli: MJP Publisher)
[22] Carpinteri A 2017 *Advanced Structural Mechanics* (Boca Raton: CRC Press)

IOP Conf. Series: Materials Science and Engineering **843** (2020) 012009 doi:10.1088/1757-899X/843/1/012009

Rebuilding the wheelset flanges of locomotives by surfacing without wheeling out

E M Martynov, V I Shkodkin and E F Romanenko

National Research Technological University «MISiS», 4, Leninsky Prospect, 119049, Moscow, Russia

E-mail: e.m.martynov@mail.ru

Abstract. The most effective method to increase the surface life of wheel tires, the flange wear of which outpaces the growth of tire wear significantly, is surfacing of worn flanges of electric locomotive wheel sets. This method, in comparison with the restoration by turning and forming a profile configuration by reducing the thickness of the tire has undeniable advantages: flange surfacing combined with the shape of female profile minimizes the technological wear of the tire before its retirement; idle time of locomotives caused by the need of intermediate turning is eliminated; the costs of maintaining the machines for turning without removing the wheels and purchasing hard-alloyed cutting tools are decreased; wage savings are achieved. The work presents basic information on wheel sets, theoretical data on surfacing, analyzes previously used surfacing materials and provides metallographic studies of the surfaced material. The technology of flange surfacing without removing the wheels from under the electric locomotive was developed and described, that allows reducing the time spent on repair significantly. The use of alloyed surfacing materials instead of low-carbon ones allows reducing the time of surfacing and increasing the resistance to wear of wheel set flanges of traction vehicles. Efficiency of NP-18H1G1M.01 surfacing wire application for wear-resistant surfacing of tire flanges and increase of the overhaul cycle is shown. Increased wear resistance of electric locomotive wheel flanges surfaced with alloyed wire is confirmed by the production test data.

1. Introduction

Intense lateral wear of rails and rolling stock wheel set flanges especially in curved tracks, is a threat to the safe movement of a traction vehicles and causes increased costs. The urgency of this issue grows as the service life of rails, flanges of locomotive wheel sets and solid wheels of wagons decreases.

When the rail track and rolling stock interact, contact stresses occur at the contact points between the wheels and the rails. As a result of these stresses, the movement of the wheels on the rails results in natural wear and tear of the mating surfaces, as well as their elastic and plastic deformation and fatigue damages.

The rules of technical operation of the railways do not allow traveling the trains with a transverse crack in any part of the axis of a wheel set, as well as in the presence of wear and damages that disturb the normal interaction between the rail track and the rolling stock.

These malfunctions include:

- wear around the taping line more than permissible dimensions;
- flange wear below the maximum permissible thickness;

- thin flange and sharp flange;
- thickness and width of wheel rim less than permissible dimensions, as well as local widening;
- defects on the running surface in the form of flat spots, shelled tread, ring wear above the permissible ones.

When the set values of the wear are achieved or the damages which threaten the safe running occurred, the wheel sets shall be taken out of service for maintenance.

2. Materials and methods

The most effective method to increase the service life of a wheel set tires of industrial electric locomotives in which the wear of the flange significantly outpaces the growth of tire wear (Figure 1), is surfacing of worn flanges without wheeling out. This method allows keeping the profile of the wheel thread and reducing the maintenance downtime of electric locomotives [1] in comparison with the restoration by turning and obtaining a profile by reducing the thickness of the tire.

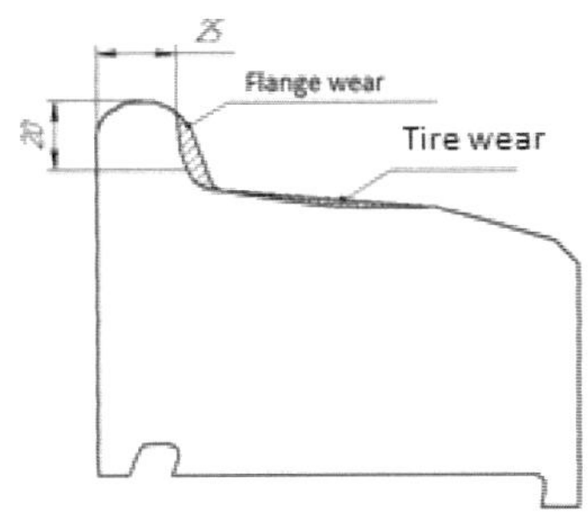

Figure 1. Flange wear

Until recent times the design documentation was developed and the unit for wheel set flanges surfacing under a flux without removing the wheels was manufactured during the restoration of the flanges of OPE and NP industrial electric locomotives at PJSC "Mikhailovsky GOK".

For surfacing the flanges Sv-08GA GOST2246-70 wire and AN-348AM flux as per GOST9087-81 were used that provided the hardness of the surfaced material surface ≈205 HB that is less than hardness of the flanges in the delivery condition 231 ... 241 HB made of steel grade 2 according to GOST 398-96 [2].

Flange restoration practice by low-alloyed wire SV-08GA under a layer of flux of AN-348AM grade has shown that the wear resistance of the surfaced material is insufficient and makes 25-35% of flange metal wear resistance.

This became the basis for a study on the use of new material to increase the hardness of the surfaced metal and increase the service life of the surfaced flanges [3-5].

Surfacing flux-cored wire PP-Np-18H1G1M.01 Technical Specification 1274-025-37392360-2012 produced by LLC "Interpro" was chosen to conduct the research. According to the certificate, the hardness in the fourth surfaced layer is 280 HB, which meets the tasks of increasing the overhaul cycle when surfacing the flanges of industrial electric locomotives flanges without removing. The hardness of surfaced layer 393 … 415 HB was obtained on testing samples during surfacing of wheel set flanges with abovementioned wire. The hardness was measured by TK-2M gauge.

To determine further prospects for the use of the wire a comparative production tests of wheel sets and flanges on railway tracks in the open mine, which were surfaced with SV-08GA and PP-Np-18H1G1M.01 wires, were carried out. Flanges were surfaced up to the thickness of 33 mm with preheating of ≈ 2650 C at flange wear up to min allowed thickness of 25 mm [6]. The heating temperature was calculated using a formula that takes into account the chemical composition and thickness of the metal. During the tests wear fixation of restored tires was done as per the time.

According to the conducted tests and collected data, the dependences of wear and tear of flanges were deduced for tested wires Sv-08GA and PP-Np-18H1G1M.01. From the composed dependencies it can be noted that the operating time up to the min allowed wear of the flanges surfaced with PP-Np-18H1G1M.01 wire is 3 times longer than that of Cv-08GA wire (Figure 2).

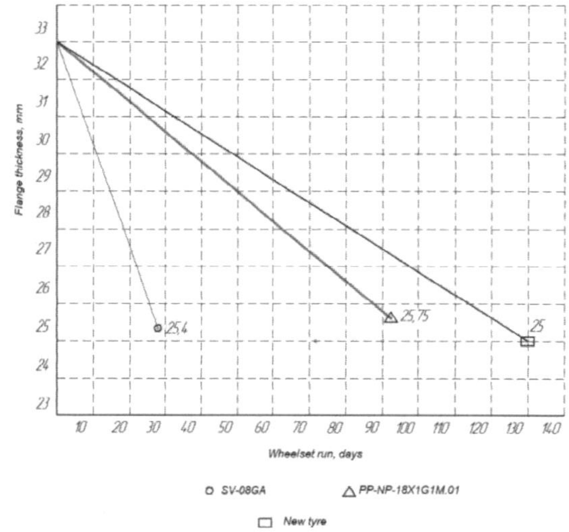

Figure 2. Wheel set flanges surfacing wear resistance diagram.

To study the macro- and microstructures of the surfaced metal, the templates (Figure 3) were made by surfacing a worn-out tire with wires Sv-08GA and PP-Np-18H1G1M.01.

Figure 3. Wheel set tire template with surfaced flange

IOP Conf. Series: Materials Science and Engineering **843** (2020) 012009 doi:10.1088/1757-899X/843/1/012009

During the study of the flange template macrostructure the cracks in the surfaced layer and heat affected zones were not revealed.

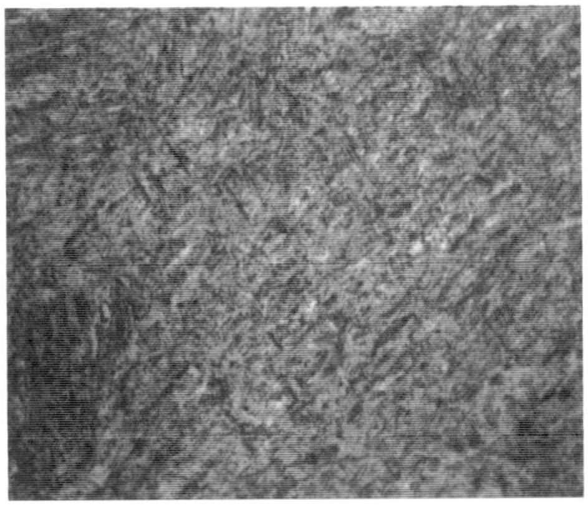

a)

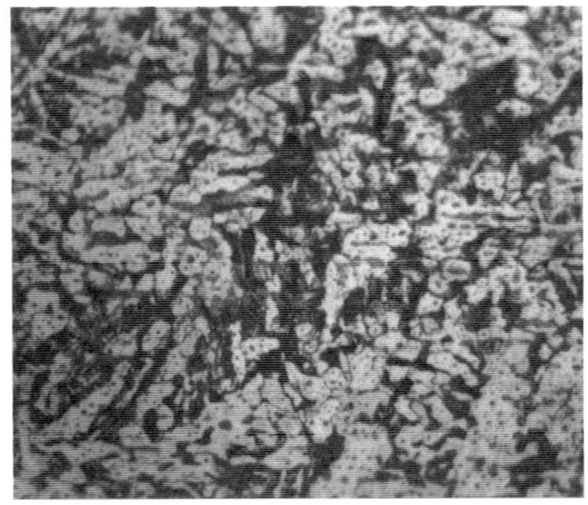

b)

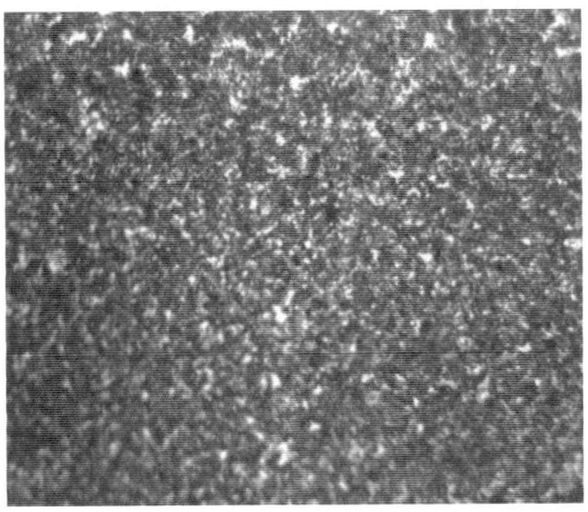

c)

d)

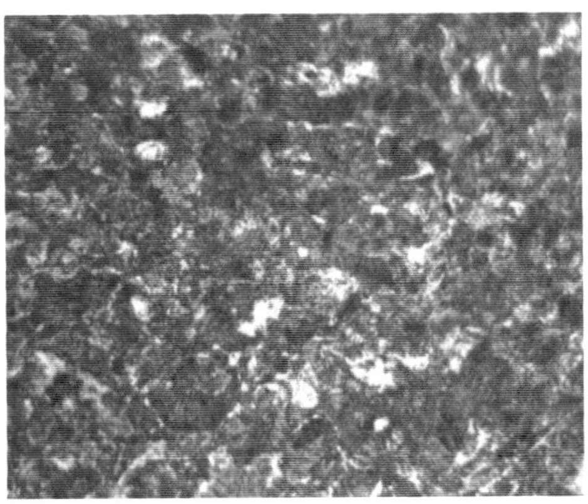

e)

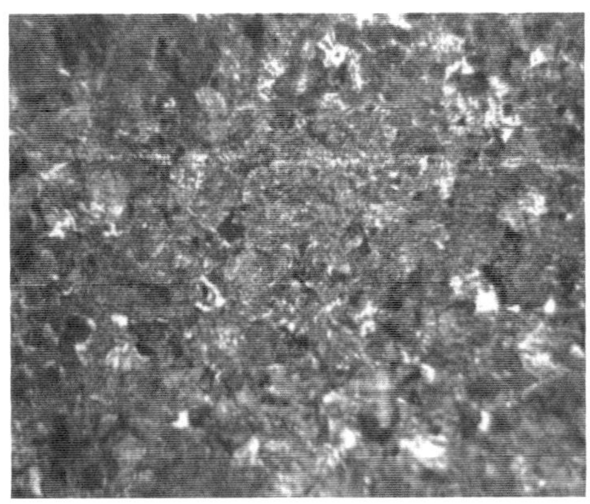

f)

Figure 4. Microstructures (×500) of surfaced and base metal of flange: a – with PP-Np-18H1G1M.01 wire , b – SV-08GA (b): c, d – 3TV; e, f – base metal.

Microstructure of the surfaced layer was analyzed with the help of metallographic optical inverted microscope METAM LV 32. The images of surfaced samples microstructures were obtained in a reflected light with resolution magnification of the microscope × 500. Micro sections of surfaced samples were prepared to high finish by preliminary abrasive grinding and three-stage polishing using ASM pastes with abrasive grain sizes of 6, 3 and 1 micron. For clearer identification of the microstructure the etching with previously prepared 15% alcohol solution of nitric acid was carried out.

The microstructure of base and surfaced metal is shown on Fig. 4. Microstructure of the tire is a finely dispersed secondary sorbite with metal hardness of 231 ... 241 HB (Fig. 4, e, f). In the microstructure of the layer surfaced with SV-08GA wire the ferrite component with hardness of the surfaced layer 205 HV prevails (Figure 4, b). During the study of the microstructure of the layer surfaced with Np-18H1G1M.01 wire, the prevalence of sorbite with the hardness of the surfaced layer of 305...415 HB is observed (Figure 4, a).

Table 1. Average value of surfaced and base metal hardness

Surfacing wire	Average hardness, HB	
	Surfaced metal	Base metal
Sv-08GA	205	241
PP-Np-18H1G1M.01	404	

From the table it follows that the metal surfaced with PP-Np-18H1G1M.01 wire has a hardness of ≈ 404 HB, which corresponds to the hardness of the rails material. As the hardness of heat-strengthened rails is 311 … 420 HB, it corresponds to the correlation of a tribo-system "wheel flange-rail" in the range of 1.0 -1.2. It is appropriate to note here that at this level of wheel hardness no accelerated wear of rails was noticed.

Hardness of metal surfaced with Sv-08GA wire, turned out to be below the hardness of flange metal surfaced with tested wire, i.e. the results of practical application of PP-Np-18H1G1M0.1 wire were confirmed.

3. Conclusion

1. The service life of wheel set flanges surfaced with PP-Np-18H1G1M.01 wire during performance tests is almost 3 times higher than those surfaced with Sv-08GA.
2. The change of the tribo-system "wheel flange – rail" hardness correlation in the range of 1.0-1.2 (for PP-Np-18H1G1M.01 wire) does not cause accelerated wear of rails.

References

[1] Buinosov A P 2010 Model of railway transport complex systems service wear *Transport annual of Volga region* **4 (24)** 21-25
[2] BuinosovA P 2009 *The main reasons for the intense wear of rolling stock wheelset tires and methods of their elimination* (Ekaterinburg: USURT)
[3] Tolstov I A, Korotkov V A 1990 *Surfacing guide* (Chelyabinsk: Metallurgy)
[4] Grokholskiy N F 1966 *Restoration of machine parts and mechanisms by welding and surfacing* (Machine engineering)
[5] Grigoriev S B, Efimenko L A, Salnikov V G and Gadalov V N 2013 Evaluation of hard - facing and process of its application on drill pipe tool joints *Chemical and Petroleum* **48** 642-645
[6] Lavrentev A Y, Dozhdelev A M, Romanenko D N and Filonovich A V 2017 Research of structural phase transformations in the fusion area of bimetallic punching tools *Journal of Chemical Technology and Metallurgy* **52** 707-710

[7] Artemenko Yu A, Emelyushin A N, Dema R R, Nefed'ev S P and Zhuravlev G M 2017 Physical modeling of the mechanism of modification with wear-resistant surfacing *Chemical and Petroleum Engineering* **52** 769-773
[8] 2001 *Industrial railway transport rolling stock safe operation and maintenance technology*
[9] Bogachev I N and Yegolayev V F 1973 *Structure and properties of iron-manganese alloys* (Mettalurgy)
[10] Novomeyskiy Yu D, Glazkov V M 1969 *High manganese austenitic steel* (Metallurgy)
[11] Schulte Yu A 1970 *Cold-resistant steels* (Metallurgy)
[12] Novomeyskiy Yu D and Lishvits V I 1964 *Properties and application of high-manganese steel* (Tomsk)

IOP Conf. Series: Materials Science and Engineering **843** (2020) 012010 doi:10.1088/1757-899X/843/1/012010

Procedure for research of process strength of deposited high-speed steel used in cutting tools

E M Martynov[1], D A Barchukov[2] and E F Romanenko[1]

[1] National Research Technological University «MISiS», 4, Leninsky Ave., Moscow, 119049, Russia
[2] Tver State Technical University, 21, Bobkova Str., Tver, 170024, Russia

Abstract. The article considers the existing research techniques applied to metal process strength. The author describes his own procedure for the research of process strength of deposited high-speed steel, taking full account of operating principal of trial plant. A quantitative criterion of the process strength of the deposited metal is proposed.

1. Introduction

Working efficiency is an object state wherein the parameter values characterizing the ability to perform specified functions meet the requirements of standards, regulations and specifications and (or) design (project) documentation. The efficiency of high-speed steel as a tool material is determined by its attribute level (mechanical, technological, operational) characterizing the ability to make a tool that meets the specified requirements.

Increasing the working efficiency and operability of tool making with the simultaneous reduction of high-speed steel consumption is a pressing problem. The solution of this problem is possible not least because of the improvement of the structural state of high-speed steels, i.e. the achievement of a duplex microstructure of tempered steel, the refinement of grains and carbides, the densification of carbon and alloying elements in a solid solution, and the martensitic hardening with high-dispersity carbides in the course of heat treatment. This solution is made possible with the introduction of deposited high-speed steel hardening into the tool manufacturing processes [1 – 4, 11-12].

However, the manufacture of deposited bimetallic tool can be associated with technological problems directly influencing its working efficiency. They results from the heat-affected zone (HAZ) occurring in a deposited bimetallic tool and the processes going on in base and deposited metals during and after metal deposition.

Strong changes in the structure and properties of a metal occur in a HAZ, which directly affects the working efficiency of a compound high-speed steel edge tool. The HAZ is extremely undesirable but, unfortunately, unrecoverable in high-speed steel deposition. However, the extent of the HAZ areas as well as their properties, including hardness can be influenced in a variety of ways [5 – 7, 13-14].

Process strength is an ability of a metal to withstand various kinds of impact without destroying in the course of its processing.

The ability to form seams without hot cracks, i.e. the ability to be in a state of elastic-plastic strain without destroying at high temperatures in the process of metal deposition or welding is called process strength of a deposited metal during the crystallization process. A quantitative assessment of

IOP Conf. Series: Materials Science and Engineering **843** (2020) 012010 doi:10.1088/1757-899X/843/1/012010

the process strength of deposited metal is basically a technique for determining the tendency of a deposited metal to form hot cracks during deposition.

There are many techniques to assess the process strength, which differ significantly in their effectiveness.

For the purpose intended, they are divided into two categories:

1. Techniques for assessing the resistance of alloys to the formation of hot cracks, regardless of the type and size of welded joints.
2. Techniques for assessing the resistance of a deposited metal to the formation of hot cracks for a given alloy composition and a deposition mode.

The classification of existing techniques for determining the process strength of welded joints is given in Table [8, 15-16].

Table 1. Techniques for assessing the resistance to the formation of hot cracks in the process of crystallization

No	Technique	Criterion	Purpose
1.	Chemistry-based calculation of the tendency of an alloy to form hot cracks	Carbon equivalent Ceq, % Amount δ – Fe, % for austenitic steels Equilibrium eutecticum E,%	Approximate quantitative assessment of alloys
2.	Deposition of laboratory technological specimens	a. Presence of hot cracks in specimen deposition b. Hot crack frequency c. Relative length of an hot crack d. Critical speed of metal deposition e. Critical size of a specimen	Qualitative assessment of alloys (a); semi-quantitative assessment (b–d) of alloys; quantitative assessment (e) of alloys
3.	Deformation of the deposited metal during its crystallization when surfacing	Critical tensile speed	Quantitative assessment of alloys.
		Critical tensile pace	Quantitative assessment of alloys and technological options of welding. Not recommended.
		Critical strain	The same.
		Critical stress.	
4.	Deposition of industry specimens	Allowable welding conditions, in which no hot crack is detected	Selection of welding technology
5.	Determination of the metal resistance margin to hot crack formation in structural welding	Difference between the actual and permissible indices of resistance to hot crack for a specified design	Assessment of the construction resistance to hot crack in welding under predetermined conditions

Most of the existing techniques are based on the application of various external forces (stretching, bending, twisting, etc.) to the specimens both during and after the metal deposition process. The results of the tests are both quantitative (speed and magnitude of strain, etc.), and qualitative indicators of process strength (presence or absence of hot cracks, etc.). The simplest scheme of inner stresses comes out of the bead deposited on a plate. For a sufficiently light plate, these stresses have a linear (uniaxial) nature; therefore their determination is reduced to the measurement of only one component of deformations [9, 17-18].

2. Experimental part

To carry out the study of the process strength of deposited high-speed steels, we used the procedure [10, 19-20] to determine the stresses in the deposited specimen in the event of hot cracks at temperatures close to the solidus curve.

When manufacturing a trial plant (Figure 1), we took into account that specimens are deposited directly during the research. Under these conditions, the values of thermal and welding strains are small, and the value of deposition strain pace does not reach values that can induce hot crack formation. So there appears a need to apply an external tensile load of an agreed value in the deposited metal brittle temperature range.

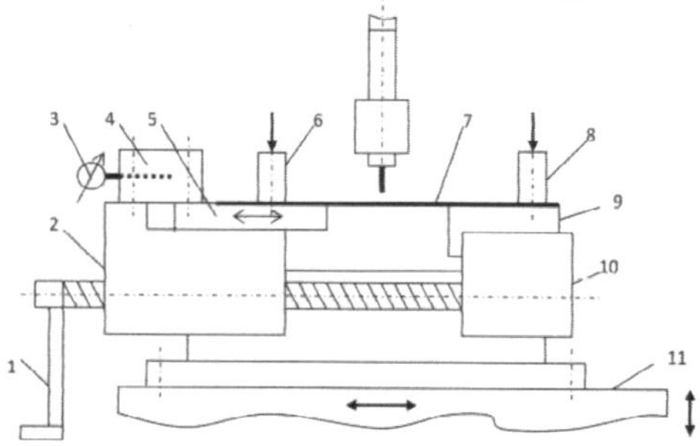

Figure 1. Trial plant for determining the process strength of a deposited metal: a - plate fixing scheme (1 - torque wrench; 2, 10 - movable and fixed jaws; 3 - indicator; 4 - spring element; 5, 9 - movable and fixed plates; 6, 8 - clamps, 7 - specimen, 11 – deposition plant table).

The trial plant has movable (due to the screw pair rotation) jaw 2 and fixed jaw 10 and is fixed on deposition plant table 11. Two plates 5 and 9 are fixed on the trial plant. Plate 5 can move freely. Spring element 4 (flat spring) designed for fixing movements and load intensity of specimen 7 during deposition is connected rigidly with a movable plate and a movable jaw of the plant.

Dial indicator 3 (with the measuring accuracy of 0.001 mm) is directly mounted and screwed down on the spring element (Figure 2). The indicator shows the movements of a spring element Δl_2 and a specimen Δl_1 subjected to tensile loading. Specimens are fixed in clamps 6 and 8 mounted on the plates. Turning a screw pair with torque wrench 1 we can preload or load during deposition specimens with the given load intensity and monitor the value of a spring element movement with an indicator.

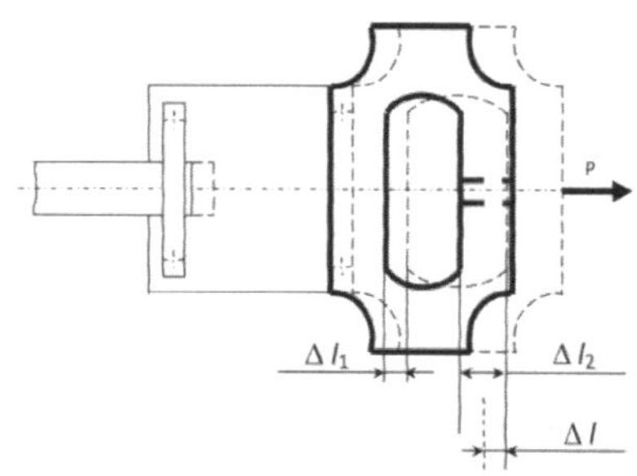

Figure 2. Movement measuring schematic: $\Delta\, l$ – indicator reading, $\Delta\, l_1$ – specimen movement, $\Delta\, l_2$–spring element movement.

A specimen form for the suggested testing of deposited metal is chosen according to the following criteria:

- A specimen must have minimum rigidity to enable testing of deposited metal at small values of tensile loading.
- Specimen dimensions must be selected so that to obtain the values of the deposited metal cooling rate not less than the critical rate of hardening.

- A specimen must provide strain concentration in the test cross-section when the specimen ends are loaded.

These criteria are met for specimens of 5×25 mm bar (steel 30HGSA (30ХГСА in Russian), $\sigma_т$=830 MPa, GOST103-76) and 200 mm length (Figure 3).

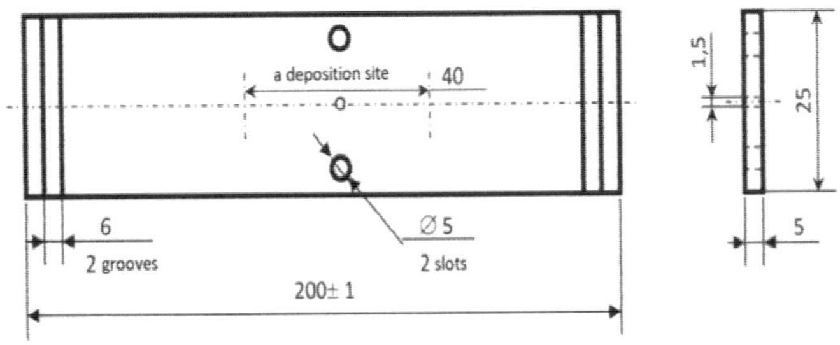

Figure 3. Test specimens to study the process strength of deposited metal.

1 mm deep grooves are milled at the ends to provide specimens with positive clamping in the rig. Two slots with 5 mm diameter are drilled in specimens to provide strain concentration in the test cross-section when the specimen is being deformed. The slot diameter is selected so that to form a quality weld bead and prevent deposited metal from leaking through slots during deposition.

1.5 mm slot is drilled in the specimen centre to fix a thermopair.

To obtain the value of tensile load the spring element is qualified which means that the dependence of spring element movement Δl_2 on the tensile load applied is determined.

The application of tensile load before and during deposition makes it possible to estimate the process strength margin of deposited metal at different temperatures in the test cross-section. The values of tensile load applied to specimens are increased consistently until hot cracks appear in deposited metal.

The criterion of the deposited metal process strength is maximal specimen strain in a brittle temperature range at the agreed value of tensile strength when hot cracks do not form in deposited metal.

The application of the same tensile load before and during deposition (in the area of elastic and plastic strain) causes various specimen deformations due to the change of base metal properties when its temperature increases.

The application of tensile load P in the area of elastic strain before deposition makes it possible to obtain the critical values of tensile stress in the test specimen cross-section which cause the formation of hot cracks in deposited metal. Indicator reading Δl, in this case, corresponds to spring element strain Δl_2.

The application of tensile load during deposition, when deposited metal temperatures in the test cross-section fit with a brittle temperature range, makes it possible to estimate the deposited metal resistance to hot crack formation with the measurement of critical specimen strain. Readings of indicator Δl correspond to the strain difference of spring element strain Δl_2 and plate Δl_1.

3. Conclusion

The prevention of crack formation is one of the main tasks in obtaining a deposited metal with alloying solid solution as hardened.

The deposition technology for manufacturing high-speed steel edge tools does not make it possible to apply fully the well-known destructive welding check, process strength testing in particular. Deposition, although being allied process to welding, is characterized by specific heat processes in base and deposited metals as well as their structural transformations.

Thus, to make deposited high-speed hardened steel with grain size 11 is a process task that can be solved successfully under a specified thermal cycle. Thereafter, the process strength criteria suggested in widely-known techniques are not appropriate in this case.

The present procedure uses the basic principles of the well-known technique determining strain rates in a deposited specimen when hot cracks form at the temperatures close to a solidus curve [10, 21] (loading diagram, the use of a spring element, tensile load during deposition, etc.).

The criterion of deposited metal process strength is a maximal specimen strain in a brittle temperature range at the agreed value of tensile strength when hot cracks do not form in deposited metal.

References

[1] Barchukov D A 2013 RU, Patent No. 2483120 Hardening method of deposited high-speed steels

[2] Barchukov D A 2012 Resource-saving technology for producing a tread cutter with a strengthened cutting part *Nauchnoye Obozrenie (Science Review)* **5** 404-407

[3] Gadalov V N, Kolmykov D V, Korenevskii N A, Chernysheva E V and Ziborova T 2010 Low-temperature nitrocementation to improve the life of resurfaced tractor crankshafts *Russian Engineering Research* **30** 1090-1091

[4] Grigoriev S B, Efimenko L A, Salnikov V G and Gadalov V N 2013 Evaluation of hard - facing and process of its application on drill pipe tool joints *Chemical and Petroleum Engineering* **48** 642-645

[5] Lavrentiev A Yu, Dozhdelev A M and Barchukov D A 2014 Change of structure and properties of the zone of thermal influence at production of the welding tool from high speed steels *Proceedings of the Southwest State University* **5 (56)** 14-20

[6] Lavrentev A Y, Dozhdelev A M, Romanenko D N and Filonovich A V 2017 Research of structural phase transformations in the fusion area of bimetallic punching tools *Journal of Chemical Technology and Metallurgy* **52** 707-710

[7] Romanenko D N, Artemenko Yu A, Emelyushin A N, Dema R R, Nefed'ev S P and Zhuravlev G M 2017 Physical modeling of the mechanism of modification with wear-resistant surfacing *Chemical and Petroleum Engineering* **52** 769-773

[8] Vinikurov V A 1979 *Welding in machine building* (Moscow: Machine Building)

[9] Nikolayev G A and Vinokurov V A 1990 *Welded Structures. Calculations and design* (Moscow: Higher School)

[10] Prokhorov N N *Hot Cracks in Welding* (Moscow: Mashgiz)

[11] Romanenko D N, Markelov E A and Nikolaenko A V 2013 Attachments for diamond smoothing *Russian Engineering Research* **33** 379-380

[12] Gorozhankin V V and Kolmykov V I 2013 Reinforcement of button roller bits by thermochemical processing *Gornyi Zhurnal* **10** 98-100

[13] Gadalov V N, Abakumov A V and Turaeva O A 2014 Effect of carbide content in carburized layers on drilling tool self-sharpening capacity *Chemical and Petroleum Engineering* **50** 547-549

[14] Kolmykov V I, Abyshev K I and Nasteka V V 2015 Effect of structural characteristics on abrasive wear resistance and impact strength of facing and carbonitrided coatings *Chemical and Petroleum Engineering* **50** 610-613

[15] Kolmykov V I, Abyshev K I, Kolmykov D V and Bedin V V 2015 Efficiency of surface hardening by carburizing steel objects operating under abrasive wear conditions *Chemical and Petroleum Engineering* **51** 58-61

[16] Shcherenkova I S, Shkatov V V and Gadalov V N 2015 Study of electrolytic chromium coatings with ultradisperse superhard fillers *Chemical and Petroleum Engineering* **51** 277-282

[17] Gadalov V N, Gvozdev A E, Vornacheva I V and Kovalev S V 2016 Burnishing with wear-resistant mineral–ceramic and hard-alloy indenters *Russian Engineering Research* **36** 731-734

[18] Shkatov V V, Pogodaev A K and Mazur I P 2017 The influence of nanoscale precipitation of aluminum nitride on the formation of recrystallization texture in aluminum deoxidized low-carbon steels *Journal of Chemical Technology and Metallurgy* **52** 617-620

[19] Romanenko D N, Kolmykov V I, Ermolaev D V and Mazur I P 2017 Increasing of fatigue limit and efficiency of constructional steels through additional chemical and thermal treating using nitrogenous carbonaceous carburizers *Journal of Chemical Technology and Metallurgy* **52** 990-995

[20] Gubanov O M, Shkatov V V and Kozhukhov A A 2017 Formation of non-uniform grain structure of steel in the process of heat treatment and method of evaluation of microstructure with significantly non-uniform grain *Journal of Chemical Technology and Metallurgy* **52** 996-1001

[21] Gadalov V N, Boldyreva O N and Kovalev S V 2018 Decreasing technological risk through optimization of maintenance of the equipment of an oil and gas complex *Chemical and Petroleum Engineering* **53** 610-613

IOP Conf. Series: Materials Science and Engineering **843** (2020) 012011 doi:10.1088/1757-899X/843/1/012011

Development of innovative approach to diagnosis of coated abrasive surface

D B Shatko, V S Lyukshin and P A Strelnikov

T.F. Gorbachev Kuzbass State Technical University, 28, Vesennyaya Str., Kemerovo, 650000

E-mail: shdb.tm@kuzstu.ru

Abstract. One of the most important performance criteria when machining parts is the surface quality, which depends on many parameters, in particular on the state of a cutting tool. However, a cutting tool eventually loses its cutting ability, and therefore, its timely diagnosis is necessary to exclude scrap in the form of a poorly machined surface. There are many methods for measuring the wear of a cutting tool, including direct and indirect ones, which allow monitoring the condition of a cutting tool and timely dressing or replacing it. At the same time, existing methods are not applicable for all types of tools, or do not provide complete information about their condition. This fully applies to a coated abrasive, the condition of which is determined, as a rule, visually or by the quality of the machined surface. Based on this, the study aimed at developing a new approach to the diagnosis of a coated abrasive face is of great relevance and practical value.

1. Introduction

Among the main methods of machining materials with a coated abrasive, the key one is grinding with a sanding belt (Figure 1), which consists of a cloth backing 1, grinding grains 2, deposited on one side of a cloth backing and bounded with an adhesive 3. Position 4 indicates a press roller, and arrows indicate the direction of rotation of a belt (*V*) and a feed (*S*).

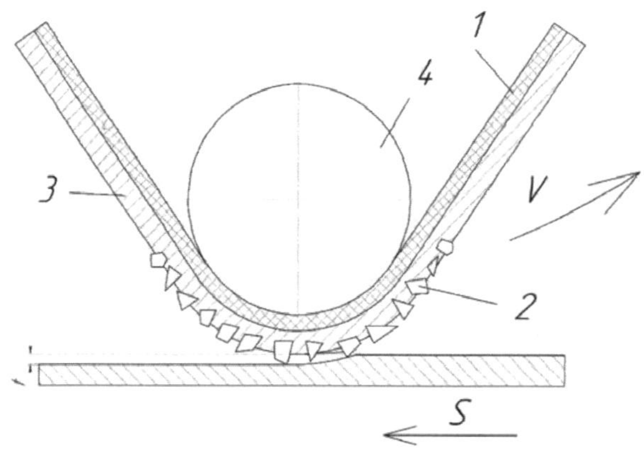

Figure 1. Belt grinding pattern

The performance of sanding belts is significantly affected by cutting elements - abrasive grains included in their composition, namely: grade, grain size, shape and orientation on the backing surface [1-5].

The currently used technology for the production of abrasive materials does not provide the uniform grain shape. The abrasive grains produced by crushing the ingot significantly differ from each other not only in terms of shape, but also in their characteristics [3].

Sanding belts, as a rule, are made with a simple pouring of the abrasive onto the backing surface, as a result of which the abrasive grains occupy a chaotic position on the belt surface. More favorable working conditions of grains are achieved by their orientation in the electrostatic field.

The orientation of grains along their major axis relative to the sanding belt surface allows the most efficient use of elongated grains ensuring their strong fixation in the bond [4, 5]. In addition to creating the optimal working conditions, the grain orientation contributes to better removal of wear products from the cutting zone.

The sanding belt surface is characterized by the following parameters: the shape and size of the abrasive, the number of grains per unit surface area, the distance between the grains and their depth. The grinding process is also influenced by the grade and uniformity of the abrasive material, as well as the properties of the bond. The listed characteristics are monitored during the manufacturing process of the tool.

One of the parameters of a sanding belt is the number of active grains per surface unit. Currently, there are theoretical and practical methods for determining the number of active grains [6-9].

In accordance with one of the existing methods, the number of active grains is determined through the equation of the lengths of the arcs of contact between the part and the grinding wheel and the equation of the number of grains acting in the contact arc. This method is relatively easy to use; however, it allows determining the number of active grains, only within the limits of contact between the grinding wheel and the part.

It is known that not all grains on the grinding wheel face are involved in the cutting process [4, 5]. According to another method, it is proposed to determine the number of active grains using a specific formula containing 8 parameters. The second method is more complex and time-consuming than the previous one, but it allows determining the number of active grains on the entire grinding wheel face.

There are other alternative methods for estimating the number of grains on the sanding belt surface using mathematical expressions to determine the average probability of the number of active (working) grains per unit surface of the contact between the belt and the part.

A practical method to accurately determine the number of active grains is foil rolling method. To assess the initial state of the sanding belt surface, it is necessary to take prints of the abrasive grains. Based on the analysis of the prints, it is possible to evaluate the parameters and the relief of the sanding belt surface.

2. Methods and research

To determine the number of active grains of a coated abrasive, a method was developed that allows counting the number of active grains using the MatLAB software. The method is as follows:

- A coated abrasive is scanned.
- The resulting digital image is processed in a software environment such as Photoshop to obtain a grayscale mode (monochrome) (Figure 2).
- The processed digital image is imported into the MatLAB software environment as a variable, which is presented in the form of a table (matrix/array).
- Based on the data fed into the MatLAB software environment, layer contour plots/surface plots are constructed.
- The graphs having been plotted; the axes limits are set at the desired depth in the editor of the properties of the axis of the graph (Figure 3).

The number of vertexes/peaks is calculated (vertexes/peaks are the active grains involved in cutting).

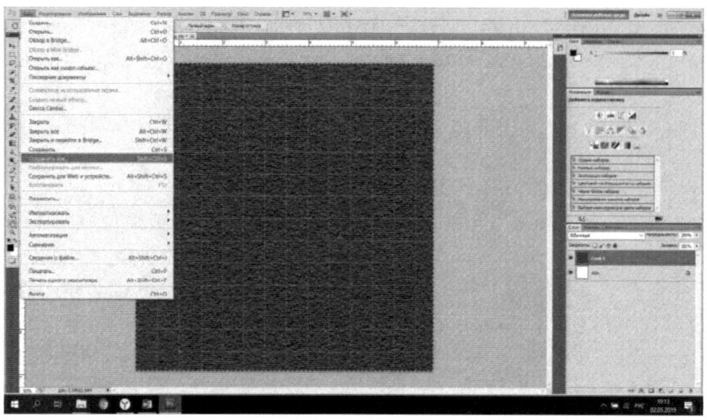

Figure 2. Digital image of the sanding belt surface

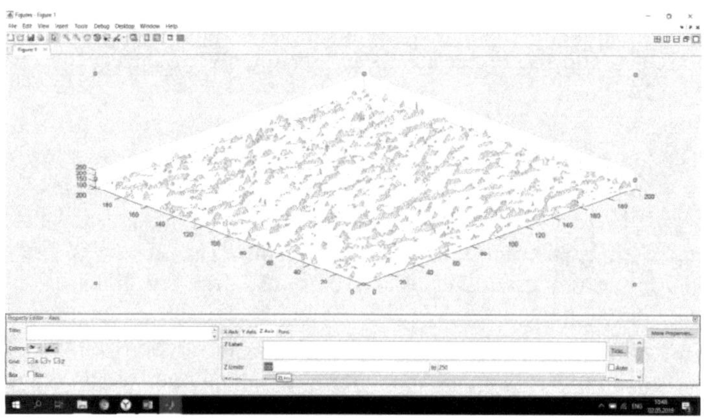

Figure 3. Resulting vertex graph

3. Results and discussion

To test the operability and adequacy of the proposed method for determining the number of active grains, it was compared with the method of foil rolling.

A coated abrasive with a size of 30x50 mm and grit No. 40 was used for this purpose. The test coated abrasive was fixed on a table and a packet of 5 carbon paper sheets with a density of 20 g/m2 was placed on top of it. Then the test coated abrasive was pressed and rolled along with a steel roller. As a result, a cast of the most prominent (active) abrasive grains was obtained (Figure 4).

Figure 4. Cast of active abrasive grains obtained by rolling

IOP Conf. Series: Materials Science and Engineering **843** (2020) 012011 doi:10.1088/1757-899X/843/1/012011

Active grains were counted manually. The rolling method showed that the number of active grains is 1233 (82 grains/cm^2).

When implementing the proposed method, the studied coated abrasive was scanned in advance (Figure 5) to obtain the smallest error in the results after rolling.

Figure 5. Scanned coated abrasive (in grayscale)

In accordance with the previously described research methodology, based on the scanned surface, a surface graph was plotted (Figure 6).

The graphs having been processed; it was estimated that, with a limit of 100 pixels, the number of grains is 1272 (84 grains/cm^2).

The number of active grains obtained by the proposed method coincides with the known method of foil rolling, so we can assume that the new method can be successfully used to determine the number of active grains.

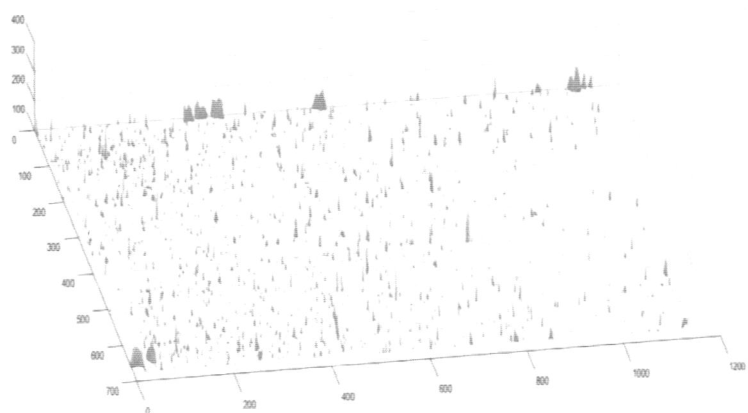

Figure 6. 100-pixel surface graph

4. Conclusion

The analysis shows that currently there are no efficient and accurate methods for monitoring a coated abrasive face. The existing methods are quite laborious, time consuming and, in addition, do not provide sufficient accuracy. The proposed approach based on the use of modern software allows avoiding the above disadvantages and can be recommended for determining the number of active grains of a grinding tool in order to predict its further operability and other performance criteria.

References

[1] Yang Y, Wang S, Geng J, Zhang J and Luo B 2019 *Beijing Linye Daxue Xuebao* **41** 138-46
[2] Syreyshchikova N and Pimenov D 2019 *J. Friction and Wear* **40** 139-45
[3] Xu Y 2017 *Jingangshi yu Moliao Moju Gongcheng* **37** 82-6
[4] Shatko D, Lyukshin V and Bakumenko V 2016 *IOP Conf. Series: Mater. Sci. Eng.* **142**
[5] Shatko D, Lyukshin V, Shutko L and Strelnikov P 2019 *IOP Conf. Series: Mater. Sci. Eng.* **142**
[6] Zou L, Liu X, Huang Y and Fei Y 2019 *Int. J. Advanced Manufacturing Technology* **104** 2961-70
[7] Li H, Li X, Chen Z, Wang L and Rong Y 2018 *Int. J. Advanced Manufacturing Technology* **97** 2125-40
[8] Verezub V 1972 *Grinding with abrasive belts* (Moscow)
[9] Zaitseva M 1966 *Abrasives and diamonds* **1** 33-7

IOP Conf. Series: Materials Science and Engineering **843** (2020) 012012 doi:10.1088/1757-899X/843/1/012012

Study on structure and mechanical properties of 20X13 steel welding joints

Y B Argirov, T M Mechkarova, A M Stoyanova, N M Atanasov, M I Konsulova-Bakalova

Technical University of Varna, Studetska Str. 1, Varna, 9010, Bulgaria

E-mail: tatuna10@abv.bg

Abstract. The article deals with the problem related to the methods of control of the structure and mechanical properties of welding joints formed during the recovery of riveted joints on turbine blades made of 20X13 steel (AISI, 420J2). For the purposes of the study, the methodologies used are for microstructural analysis, micro- and macro-hardness of two types of samples: one without further heat treatment and the other with heat treatment after the process of welding

1. Introduction

Turbines are part of a facility which can be used in many ways: as a power plant turbine (in nuclear power plants, thermal power plants, hydroelectric power plants), as a fragment of propulsion mechanisms for various modes of transport, and also as an integral part of hydraulic pumps and gas turbine engines [1].

The turbine system itself (the machine) deals with conversion of energies: internal and / or kinetic energies. It supplies the working body and allows the shaft to fulfill its mechanical purpose. The blades are the major part of this system and they are subjected to a jet of working fluid which is focused around the periphery of the rotors [2, 3].

The design of the steam turbine can briefly be described as follows. On the main element, i.e. on the shaft, a disc is fixed to which the blades are attached. Other parts, such as tube nozzles are located near these elements. Through them, steam is supplied from the boiler. As steam passes through the nozzle, it exerts a certain pressure on the blades as well as on the disk and the entire installation. The energy of the steam passing through each blade of each disc gives off some of its energy to these elements. It is namely this effect that causes the turbine disc to rotate along with the blades (Figure 1).

The important condition is the need to direct the steam jet in such a way that it does not damage the blades by its impact. This condition is possible to satisfy only with a certain shape of the surface. Through tests and calculations, it has been found that the best surface for working with steam jets is one that can provide a smooth turn, after which the movement of the working substance will be reversed in the opposite direction from the original one. In other words, it is necessary to shape the blades in a semicircle. In such a case, when faced with an obstacle, the maximum kinetic energy will be transferred to the mechanical device, which will cause it to rotate. Losses will be minimized. However, there is a noticeable wear and tear on the blades after a long period of operation[1, 3, 4].

Figure 1. Welding joints on turbine blades

Therefore, the purpose of our study are the blades of a steam turbine made of 20X13 steel worn during its operation and restored by gas tungsten arc welding (GTAW), and the subsequent riveting to the turbine shroud [5, 6].

2. Nature of analysis

As a result of emerging problems with the quality of welds (cracks and pores) in the restorative repair of steam turbine blades made of 20X13 steel, through GTAW with solid wire OK Tigrod 19.85, it is necessary to search for the causes leading to them, using methodologies for macro- and micro-structural analysis [7].

In order to carry out the study, samples were prepared of steel 20X13 (AISI, 420J2), a ferrite-martensite grade, welded by GTAW with a solid wire OK Tigrod 19.85 (BDS EN ISO 18274: S Ni 6082 (NiCr20Mn2Nb)) made by ESAB company. Its chemical composition and mechanical properties are shown in Tables 1 and 2 for the base metal, and in Tables 3 and 4 for the weld metal [7, 8, 9].

Table 1. Chemical composition of 20X13 [10]

C	Si	Mn	Ni	S	P	Cr	Fe
0.16 – 0.25	max 0.6	max 0.6	max 0.6	max 0.025	max 0.03	12 - 14	~84

Table 2. Mechanical properties of 20X13 [10]

Parameters	Tensile strength, Rm, MPa	Yield strength, Re, MPa	Total elongation A, %	Reduction of area Z, %	Impact toughness KCU, кJ /cm^2
Steel: 20H13	650-830	440-635	50-55	10-16	590-780

Table 3. Chemical composition of the material for the welding seam [8, 9]

Material	Ni	Cr	Mn	Nb+Ta	C	Si	Fe	Ti
OK Tigrod 19.85	>67	20	3	2.5	<0.5	<0.5	<1.5	<0.4

Table 4. Mechanical properties of the welding seam at 20°C [8, 9]

Electrode	Protective gas, I1	Tensile strength, Rm, MPa	Yield strength, Re, MPa	Total elongation A, %	Impact energy KV(+20), J/°C
OK Tigrod 19.85	Ar	700	425	44	100

Two samples were made of this steel for the purposes of the study. The samples were GTAW with an OK Tigrod 19.85 electrode, after which the first sample was subjected to further heat treatment, annealing at 730°C with a retention time of 30 min., whereas the second sample was left without heat treatment, as is the actual welding of the turbine blades[11, 12].

In order to test the samples, polished sections were prepared [13, 14, 15], consisting in successive polishing with a water sandpaper of different grain size and subsequent corroding with a mixed acid solution of nitric and hydrochloric acid in a ratio of 3 to 1 [16]. The observation of the macro- and micro-structure at different magnifications was carried out by a metallographic microscope NEOPHOT 2 with a Haneman 100 attachment to it [17]. The Vickers method (HV0.05 and HV5) [18] was used to measure the hardnesses in depth and in the different welding zones of the sample. The analysis was also assisted by the use of EUROMEX stereomicroscope for macro examination.

3. Investigated zones of the microsections

In Figures 2a and 2b show the photos of the polished sections with the distribution of the main zones obtained after GTAW with an OK Tigrod 19.85 electrode in both samples. Figure 2a shows Sample 1 subjected to further heat treatment, annealing at 730°C with a retention time of 30 min., and Figure 2 b) shows Sample 2 without such heat treatment [6].

In sample 1, five clearly expressed zones are visible: weld metal zone - Z.1; alloying zone - Z.2; heat affected zone - Z.3 and Z.4; The zone of base metal, where Z.3 is the precrystallization zone, and Z.4 is recrystallization zone.

In sample 2, red arrows show the four zones: weld metal zone - Z.1; alloying zone - Z.2; heat affected zone – Z.3; and base metal zone - Z.4, with Z.2.1 being the beginning of the alloying zone and Z.2.2 the end of the alloying zone. In Figure 2 is also clearly visible the imprint zone in which the HV5 macrohardness [18] was measured in both samples. For better visualization of this zone, in Sample 2, the one with no heat treatment, the zone is shown in Figure 2b enlarged in the lower right-hand corner [17].

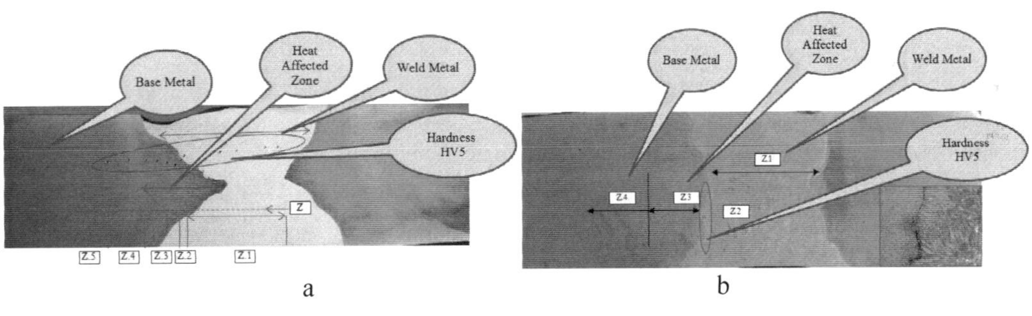

Figure 2. Investigated zones in GTAW with electrode OK Tigrod19.85: a – heat treatment; b – no heat treatment.

4. Microstructure

From the metallographic analysis thus made of the structures by help of NEOPHOT 2 metallographic microscope [17], images were obtained from the different zones of the samples. Figure 3 shows the microstructures of the two tested samples of the prepared polished sections at magnification 100 times, and Figure 4 shows the microstructures of the same two samples in the same zones as Figure 3 but at a larger magnification.

IOP Conf. Series: Materials Science and Engineering **843** (2020) 012012 doi:10.1088/1757-899X/843/1/012012

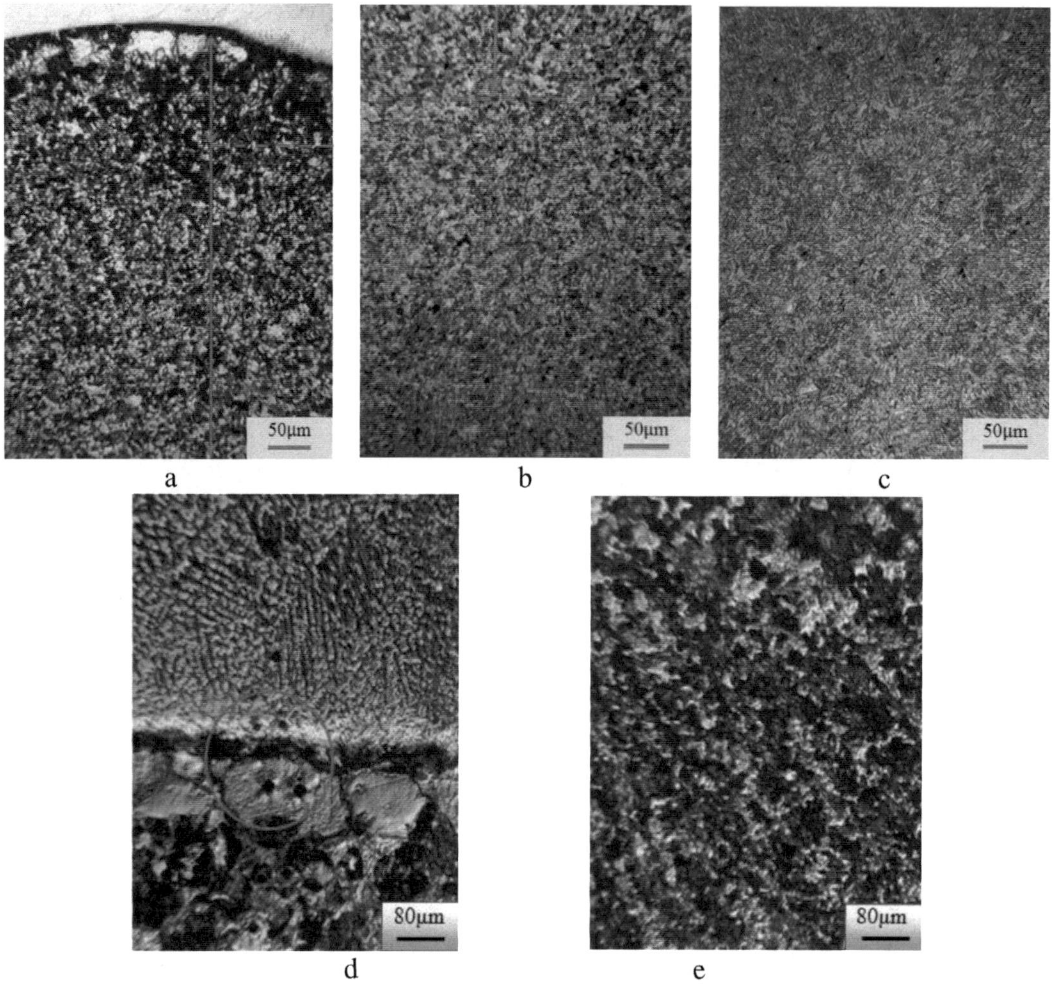

Figure 3. Microstructure of welding joint at magnification 100 times: For Sample 1 a – weld metal zone, alloying zone, heat affected zone; b – heat affected zone; c – base metal zone; For Sample 2 d – alloying zone, e – heat affected zone.

Among the considered microstructures, the following zones are of particular interest: the weld metal zones, the alloying zones and the heat affected zone, which is in contact with the alloying zone. In Sample 2, the one with no heat treatment, the alloying zone which is characterized as the zone between the liquidus and solidus lines, where the metal is partially melted, coarse grains are observed and a strongly expressed non-homogeneity with respect to the alloying zone: Figure 3 d and Figure 4 d, e. A sharp increase in hardness is observed in zone Z.2.1. This peak of increased hardness is also registered graphically when measuring the microhardness. In sample 1, after the conducted heat treatment, a homogenization of the structure is observed in the alloying zone and in the heat affected zone. The transition is more smooth in terms of the grain size (Figure 3 a, b) and (Figure 4 a, b). The weld metal zone has a dendritic austenite structure. The heat affected zones are shown in (Figure 3b and Figure 4 b) for Sample 1 after heat treatment and represent equiaxed grains with a low quantity of secondary separations, and for sample 2 (Figure 3 e and Figure 4 f), the structures have expressed secondary separations and hereditary stripeness from the base material.

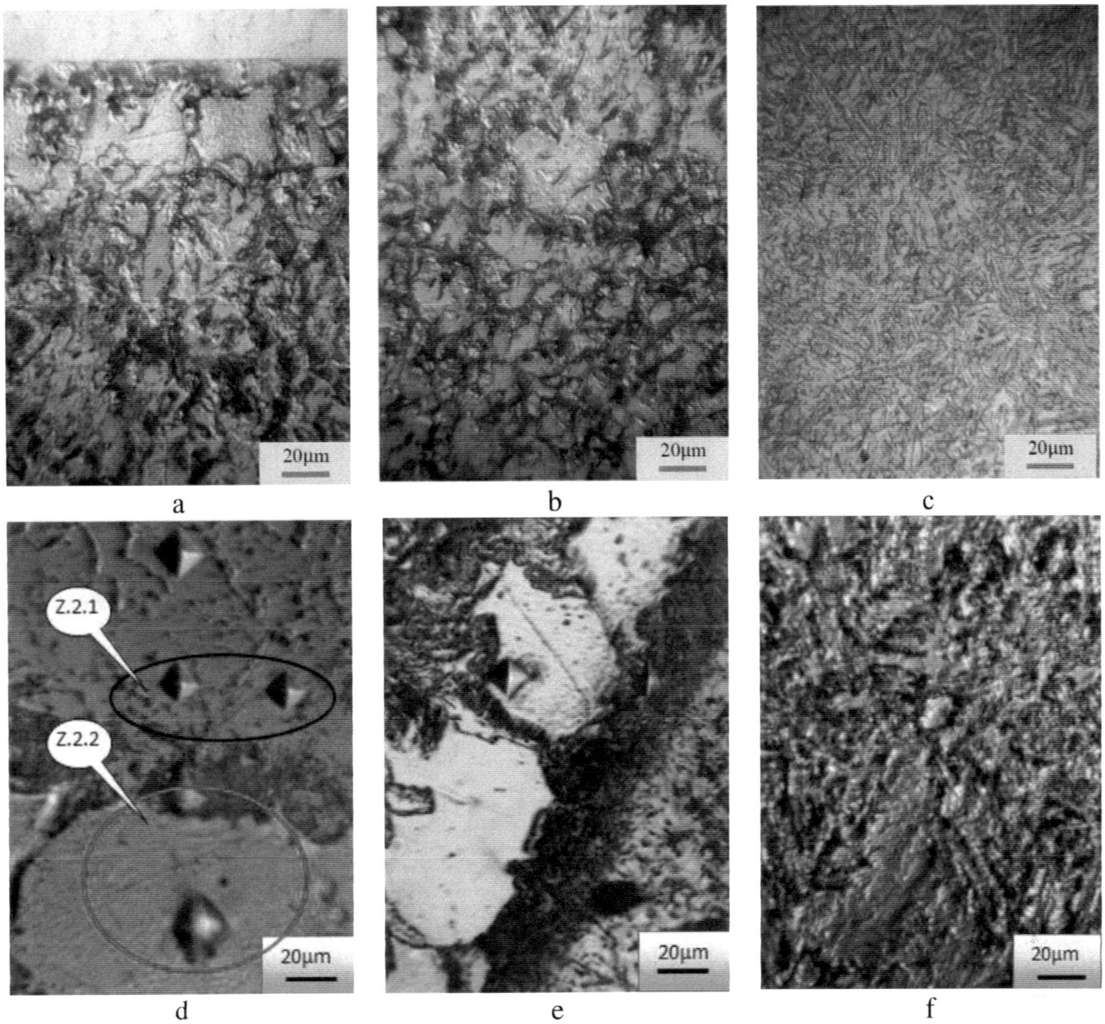

Figure 4. Microstructure of a welding joint, For Sample 1, at magnification 500 times:
a – weld metal zone, alloying zone, heat affected zone; b – heat affected zone; c – base metal zone; For Sample 2 with magnification 400 times: d – alloying zone, e – heat affected zone; f – base metal zone.

5. Macro- and micro-hardness by Vickers

After the conducted metallographic analysis of the structure, the macro- and microhardness was examined in the different zones of the samples using a NEOPHOT 2 metallographic microscope [17] and a Haneman 100 attachment to it. Macro- and micro-hardness values are registered in depth of the samples in the Z direction. The results are processed and diagrams are built as shown in Figure 5 for micro-hardness HV0.05, and in Figure 6 for macro-hardness in the welding joint HV5 [18].

In Figure 5 a, a red line shows the change in micro-hardness HV0.05 in the Z direction for Sample 1 with heat treatment, and with a blue line - the change in micro-hardness for Sample 2 with no heat treatment, as shown in Figure 5 b. The diagrams show that Sample 2 in Z.2 has a large peak of micro-hardness change, which is not favorable for the samples. And after heat treatment of Sample 1, annealing at 730°C with a retention time of 30 min., there is no such peak, which proves the necessity of this kind of processing for the turbine blades restored by GTAW [12] and subsequent riveting to the turbine shroud. In Figure 6, the same peak in Z.2 in Sample 2 is observed from the distribution of macro-hardness in the two samples, which proves the claim made for Figure 5.

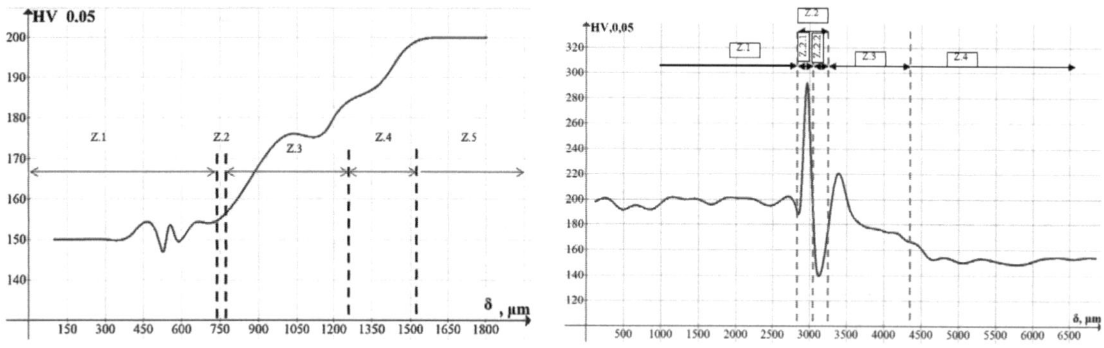

a b

Figure 5. Distribution of microhardness HV0.05 in the welding joint in the Z direction:
a - Sample 1; b - Sample 2

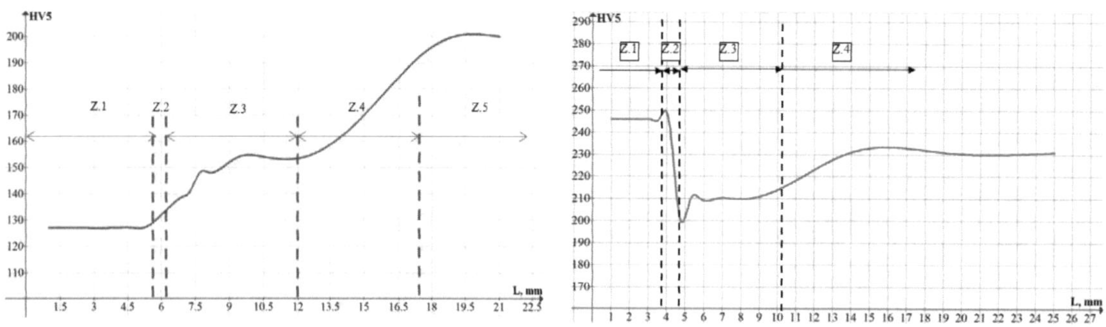

a b

Figure 6. Macro-hardness distribution in HV5 welding joint: a - Sample 1; b - Sample 2

6. X-ray analysis of the samples

X-ray phase analysis was performed for the phase transformations in both samples using a DRON-4C X-ray diffractometer. This method is universal for investigation of materials. Through it, quantitative and qualitative analyses of complex materials can be carried out.

Figure 7 shows the radiographs of Sample 1, and Figure 8 - of Sample 2 in the different welding zones.

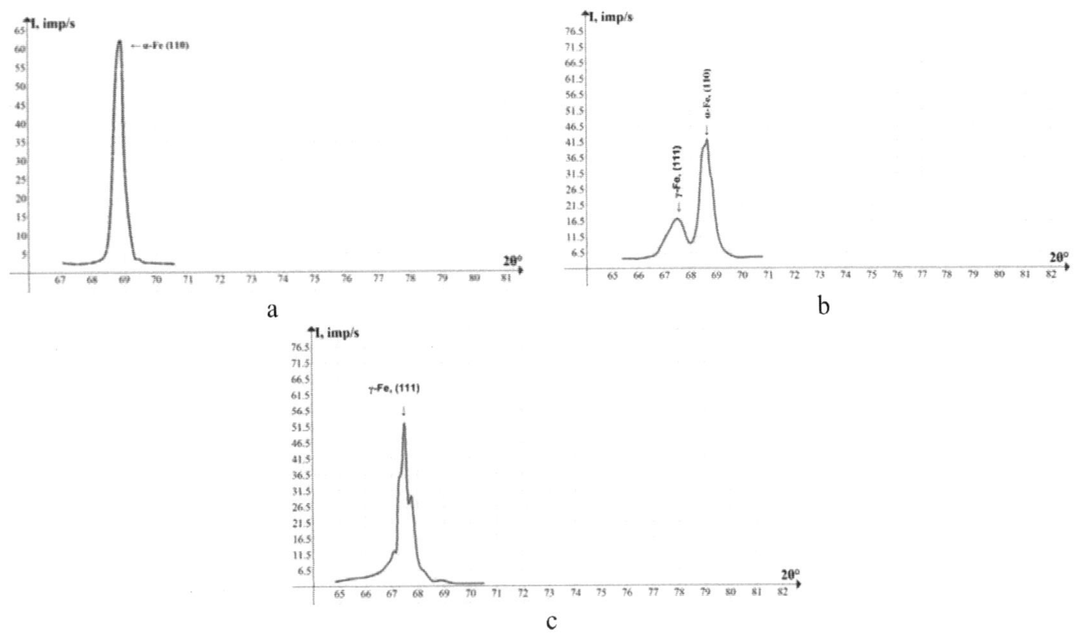

Figure 7. Radiographs of phase transformations in Sample 1: a - initial material; b - heat affected zone; c - weld metal zone

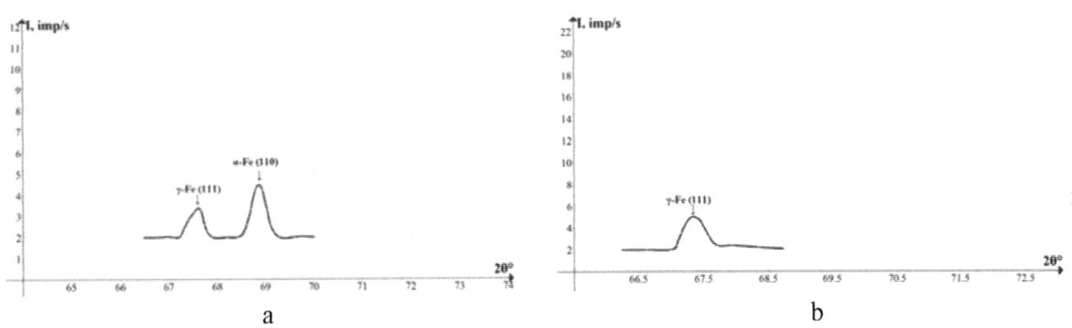

Figure 8. Radiographs of the phase transformations in Sample 2: a - heat affected zone; b - weld metal zone

The X-ray structural analysis thus carried out in the investigated zone on both samples has shown the presence of duplicity in the zone (austenite and ferrite). In the heat-treated Sample 1, the austenite participation is significantly reduced (37%, γFe -Figure 7 b), in Sample 2 with no heat treatment the quantities of the two phases are equal (48%, γFe -Figure 8 b). From the X-ray structural analysis thus carried out (Figure 8 c) in direction of the weld metal, a clearly expressed austenite structure is observed from which secondary separations are observed (Figure 4d, f) between the zones (Z.2.2 and Z.2.1.). In the weld metal zone, a well-expressed peak (Figure 7 c) is visible with a low degree of expansion compared to sample 2 (Figure 8 c), where the X-ray profile has expanded significantly and is a criterion for increased values of internal stresses.

7. Conclusion

From the conducted structural tests, the data was collected about the following zones and structures after welding: the alloying zone (the zone of partial melting of the base metal), the precrystallization zone (Ac_1, 820°C-Ac_3, 950°C), the heat affected zone including the two zones and the zone of the base metal in the investigated samples.

From the macrostructural tests, the individual zones characteristic of the welding joints have been observed and shown in Figures 4 and 5. The figures provide information on the purity of the welding joint, lack of defects and non-metallic inclusions. The zone of the weld metal, which was welded with the selected electrode, has a homogeneous austenite structure due to the high nickel content in the electrode.

It can be noted that in the alloying zone, there is coarsening of the grains in a very thin segment of about 20μm. This coarsening is only in the protruding parts of the joint, the root of the weld seam – Figure 2; Figure 3a; Figure 4a.

The structure in the precrystallization zone in the overheating zone is fine-grained, which is an indication of good mechanical properties, Figure 3b; Figure 4b. Carbides of the $M_{23}C_6$ type are observed along the grain boundaries.

In the base metal zone, Figure 3c; Figure 4c, a sorbite-troostite structure is observed.

From the conducted mechanical tests of micro and macro hardness, Figure 5, Figure 6, changes are observed in the hardness in the individual zone of the welding seam in both samples. It can be noted that in the heat affected zone there is no sudden increase in hardness, but a gradual increase towards the base material.

References

[1] Grujicic M, Arakere G, Subramanian E, Sellappan V, Vallejo A and Ozen M 2010 Structural-response analysis, fatigue-life prediction, and material selection for 1 MW horizontal-axis wind-turbine blades *J Mater Eng Perform* **19** 790–801

[2] Jensen P H 1986 *Static test of wind turbine blades. Test station for windmills* (Roskilde, Denmark: Riso National Laboratory)

[3] Jensen P H, Krogsgaard J, Lundsager P, Rasmussen F 1986 Fatigue testing of wind turbine blades *EWEA conferences and exhibition* (Rome, Italy)

[4] Dutton A G, Bonnet P A, Hogg P, Lleong Y L 2010 Novel materials and modelling for large wind turbine blades*Proc Inst Mech Eng Part A J Power Energy* **224** 203–10

[5] Bai J W, Liu P P, Zhu Y M et al. 2013 Coherent precipitation of copper in Super 304H austenite steel *J Mater Sci Eng A* **584** 57–62

[6] Spasova D 2019 Investigation of the abrasion resistance of stainless-steel composites with Al2O3 reinforcement phase produced by using the capillary forming method *IOP Conference Series: Materials Science and Engineering* 012035

[7] Product Catalogue ESAB, 3rd edition *Standard Equipment and Welding Automation*, available at: https://www.esab.bg/

[8] Product Catalogue ESAB, *Manual Metal Arc Electrodes catalogue on Direct Industry*, available at: https://www.esab.bg/

[9] Product Catalogue ESAB, Technical Handbook, *Stainless Steel Welding*, available at: https://www.esab.bg/

[10] *Steel Classification according to EN 10027-2., Designation systems for steels,* available at: http://www.steelnumber.com/en/steel_composition_eu.php?name_id=81.

[11] Spasova D 2016 Capillary Casting of Iron-Based Alloys *Advances in Materials and Processing Technologies* **2 (03)** 361 – 366

[12] Manabe Y, Wada H, Zenitani S, Hiromoto Y and Hashimoto Y 2000 Investigation on TIG welding using 2 filler wires with electromagnetically controlled molten pool process in horizontal position *Quarterly Journal of the Japan Welding Society* **18 (1)** 40-50

[13] Ueguri S, Tabata Y, Shimizu T and Mizuno T 1986 A study on control of deposition rate in hot-wire TIG welding *Quarterly Journal of the Japan Welding Society* **4 (4)** 678-684

[14] Bakalov I 2014 Methods and instruments for measuring torque and speed of marine diesel engines *Machines Technologies Materials* **7** 26 - 28

[15] Bakalov I 2015 A testing station for studying the combustion of fuel with mechanical centrifugal and rotating cup burners *Machines Technologies Materials* **8** 11 – 13
[16] Valiente Bermejo M A 2012 Henry Granjon Prize Competition 2012 Winner Category B: Materials Behaviour and Weldability Influence of the [Creq + Nieq] alloy level on the transition between solidification modes in austenitic stainless steel weld metal *J Weld World* **56** 2–14
[17] Hardmetals - Metallographic determination of microstructure - Part 1: Photomicrographs and description (ISO 4499-1:2008)
[18] Metallic materials - Vickers hardness test - Part 1: Test method (ISO 6507-1:2018).

IOP Conf. Series: Materials Science and Engineering **843** (2020) 012013 doi:10.1088/1757-899X/843/1/012013

Study of structure and physico-mechanical properties of welding joints on vessel tank of austenite steel SS316

A M Stoyanova, T M Mechkarova, Y B Argirov, M I Konsulova-Bakalova and N M Atanasov

Technical University of Varna, Studetska Str. 1, Varna, 9010, Bulgaria

E-mail: tatuna10@abv.bg

Abstract. The article addresses some problems related to the occurrence of cracks and defects in welding joints on a horizontal steel vessel tank after machine gas metal arc welding (GMAW). Tests have been carried out on the structure and mechanical properties of the welding joints between the cylinder and elliptical bottoms made of austenite steel SS316 according to AISI (X5CrNiMo17-12-2 according to BDS EN 10088-2). Proven methodologies of microstructural and X-ray analysis were used to investigate the changes in the structure (weld seam, heat affected zone and base metal). The mechanical characteristics are determined by measurements of macro- and micro-hardness by Vickers and of static tension.

1. Introduction

Radiographic inspection of welding joints after machine gas metal arc welding (GMAW) of a vessel tank [1] of austenitic SS316 steel reveals some cracks. The vessel tank is a horizontal cylindrical vessel closed on both sides with elliptical bottoms of diameter D=1.2m and length L=2.0m (Figure 1). The hull thickness is 3mm. The vessel tank is supported by two saddle posts welded to the cylinder.

Figure 1. 3D model of a vessel tank made of austenite steel SS316

The elements of the vessel tank [2] that are in contact with the working fluid in it are made of austenite steel SS316, and the saddle supports are made of carbon steel. Upon completion of the welding, they are primed with a zinc-rich primer. The vessel [3, 4, 5] is externally insulated from the environment by thermal insulation 100mm thick, since the fluid stored therein is water at a temperature of 7°C [5].

2. Essence of analysis

In order to determine the causes of cracks and defects in the welding joints, samples of austenite steel SS316 are prepared. The chemical composition of the steel (SS316) used for the purposes of this study is shown in Table 1[6, 7].

Table 1. Chemical composition of SS316 [6]

C	Si	Mn	P	S	Cr	Ni	Mo	N
0,07	1.00	2.00	0.05	0.03	16.5-18.5	10.0 - 13.0	2.0 - 2.5	0.11

The GMAW of the samples is carried out with ESAB OK Autrod 316LSi stainless steel wire BDS EN ISO14343-A: G 19 12 3 LSiSFA/AWS 5.9: ER 316LSi [8, 9] with diameter of the wire Ø0.8mm. This wire has good resistance to intercrystal corrosion and general corrosion in wet environment up to a temperature of 400°C, as well as resistance to weld scale formation up to 800°C [6, 8, 10].

The typical chemical composition and mechanical characteristics of the metal of the weld seam are shown in Table 2 [11].

Table 2. Chemical composition and mechanical properties of the metal of the seam

C	Si	Mn	Cr	Ni	Mo	Tensile strength, Rm MPa	Yield strength Re MPa	Total elongation A_5 %	Impact energy KV, J/°C		
									+20	-60	-196
<0.03	0.80	1.90	19.0	12.0	2.70	620	440	37	120	95	55

Welding is performed with a semi-automatic welding machine in the following modes [12]:

- DC and polarity (+);
- Protective gas M13 ~ 96% Ar (argon) + 4% O_2 (oxygen).
- Protective gas consumption - Q_w = 12 [l/min];
- Current- I=55-160 [A];
- Voltage- U=12-24 [V];
- Welding velocity- V_w=4-17 [m/min];
- Wire feed velocity- V_{wf}=15 [m/min];
- Melting performance- H=1,0-4,1 [kg/h].

After the GMAWof the cylinder to the elliptical bottoms, two samples of the problem areas are cut out: Sample I and Sample II as is shown in Figure 2.

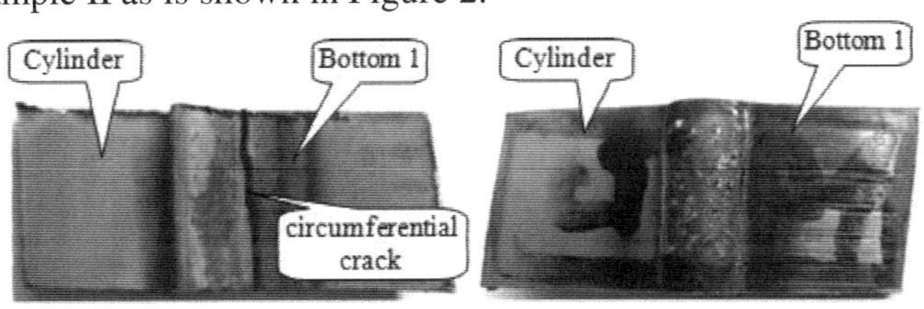

Inside view Outside view

a

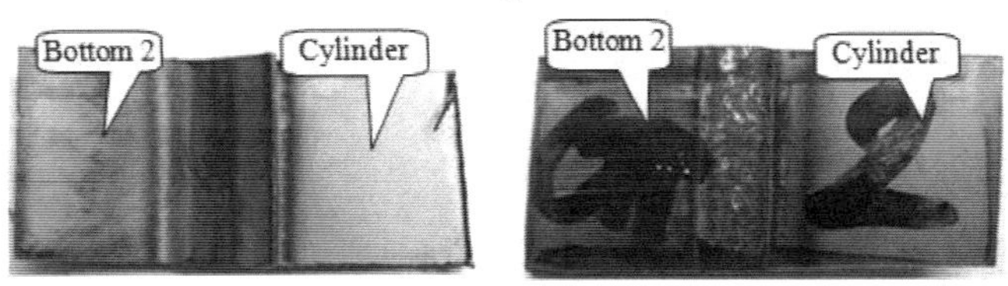

Inside view Outside view

b

Figure 2. Macro images of the cut samples for testing: a- Sample I; b – Sample II.

For the purposes of this study, the samples were polished [13] with a water sandpaper, then the polished surfaces were treated with an acid solution between nitric and hydrochloric acid in a ratio of 3 to 1. A metallographic microscope NEOPHOT 2 [13] was used for observation with different magnifications of the macro and microstructure with a device Haneman 100 attached to it. The Vickers method (HV0.05 and HV5) [14] was used to measure the hardness of the samples in the different welding zones. The phase composition of the samples was determined with X-ray apparatus model DRON 4. A stereo-microscope for macro-examination EUROMEX was also used in the analysis.

3. Macrostructure of polished samples

Figure 3a and 3b show macro images [13] of the polished samples with notations of the different zones in the structure obtained after GMAW with the OK Autrod 316LSi electrode.

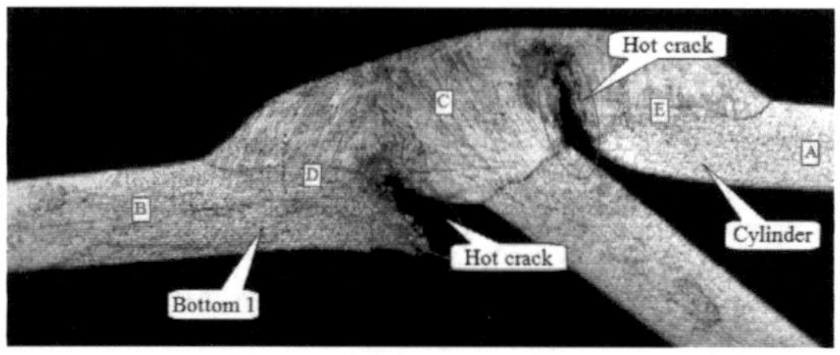

a

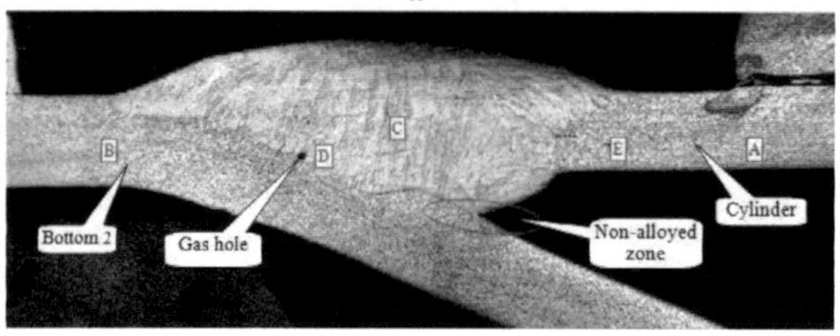

b

Figure 3. Macro-structure of the polished samples after GMAW with OK Autrod 316LSi electrode: a – Sample I; b – Sample II.

4. Microstructure of polished samples

From the metallographic analysis of the structures using a NEOPHOT 2 metallographic microscope [13], images were obtained from the different zones of the samples. Figures 4 and 5 show the microstructures of the two samples at magnification 400 times and the locations with the measured microhardness HV0.05 in depth of the samples.

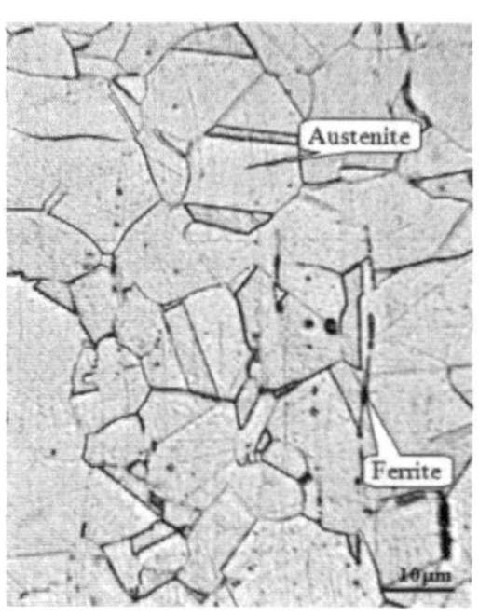

a

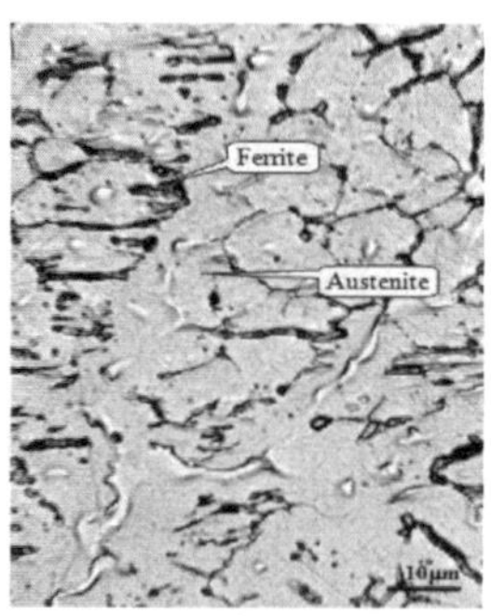

b

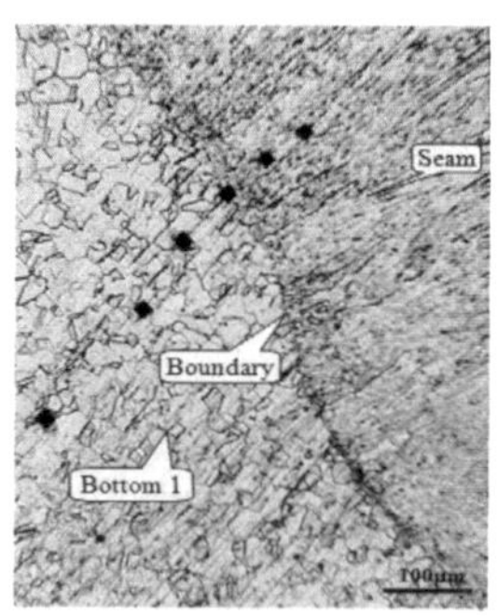

c

Figure 4. Microstructure of welding joint at magnification 400 times for Sample I: a – Base Meral Zone; b – Weld Metal Zone; c – Alloying zone.

IOP Conf. Series: Materials Science and Engineering **843** (2020) 012013 doi:10.1088/1757-899X/843/1/012013

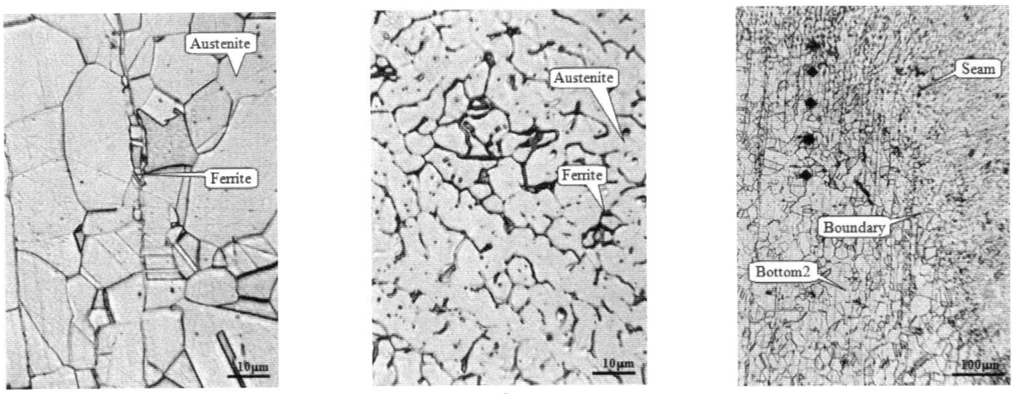

a b c

Figure 5. Microstructure of welding joint at magnification 400 times for Sample II: a – Base Meral Zone; b – Weld Metal Zone; c – Alloying zone.

The microstructural analysis of the different zones in the two samples reveals that the structure of the base metal and the weld seam material is austenitic with traces of textured ferrite. Deformation stripeness is registered in the inner zones. The visual control and microscope observation of the weld seam material do not detect any gas or non-metal defects caused by the welding.

5. Macro-hardness by Vickers

After the metallographic analysis of the structure, macro-hardness [14] was investigated in the different zones of the samples, using the NEOPHOT 2 metallographic microscope for this purpose. The values were recorded in the depth of the samples and are digitally indicated in Figure 6-a, for Sample 1 and in Figure 6-b for Sample 2.

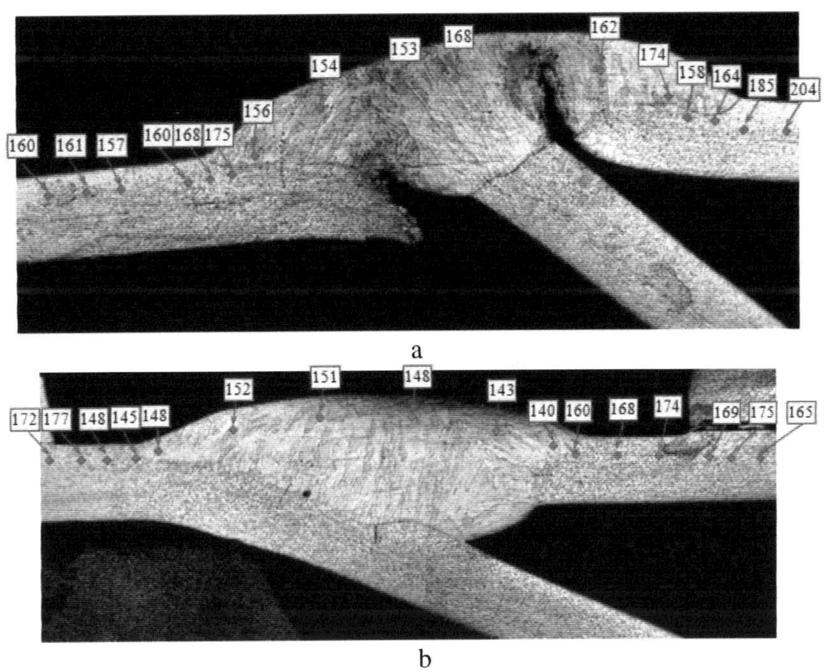

b

Figure 6. Distribution of HV5 macro-hardness in the different zones of the welding joint from left to right: a – Sample I; b – Sample II.

IOP Conf. Series: Materials Science and Engineering **843** (2020) 012013 doi:10.1088/1757-899X/843/1/012013

6. Micro-hardness by Vickers.

Figures 7 and 9 show the locations and the recorded value of the measured Vickers micro-hardness HV0.05 [14]. The figures show that lower or the same hardness is registered in the weld seam material relative to the hardness of the base material.

Thus, to better illustrate the results obtained from the measured micro-hardnesses HV0.05 in two samples, graphical dependencies are shown in Figure 8.

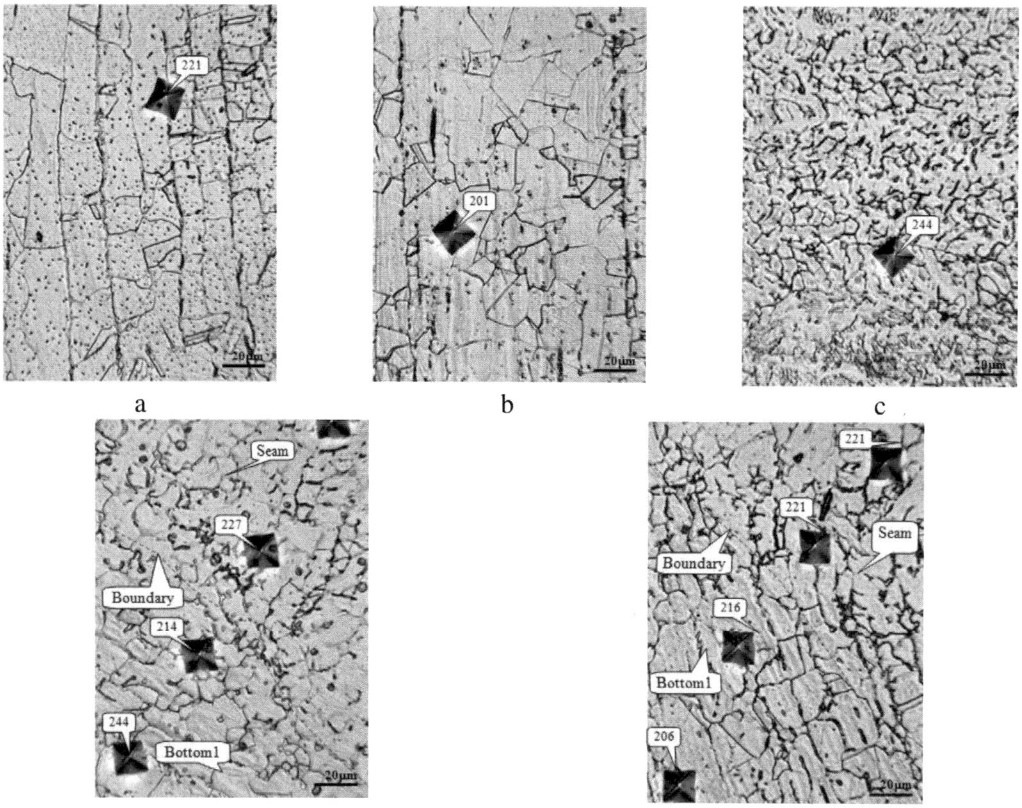

Figure 7. Micro-hardness HV0.05 for Sample 1: a and b base metal zone; c - weld metal zone; d - and e - alloying zone

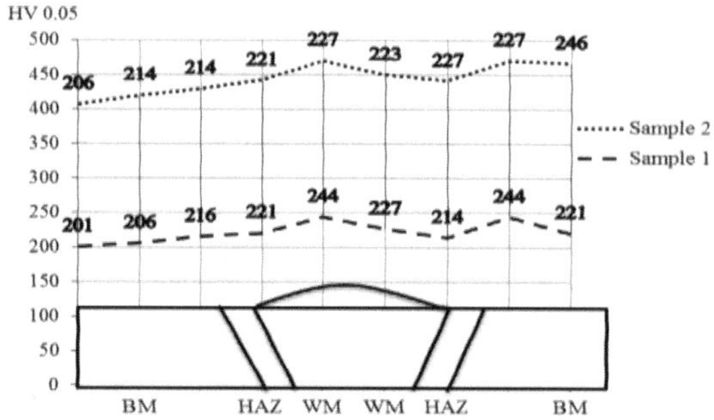

Figure 8. Graphical representation of the results from the measured micro-hardness HV0.05 for Sample I and Sample II in the different zones.

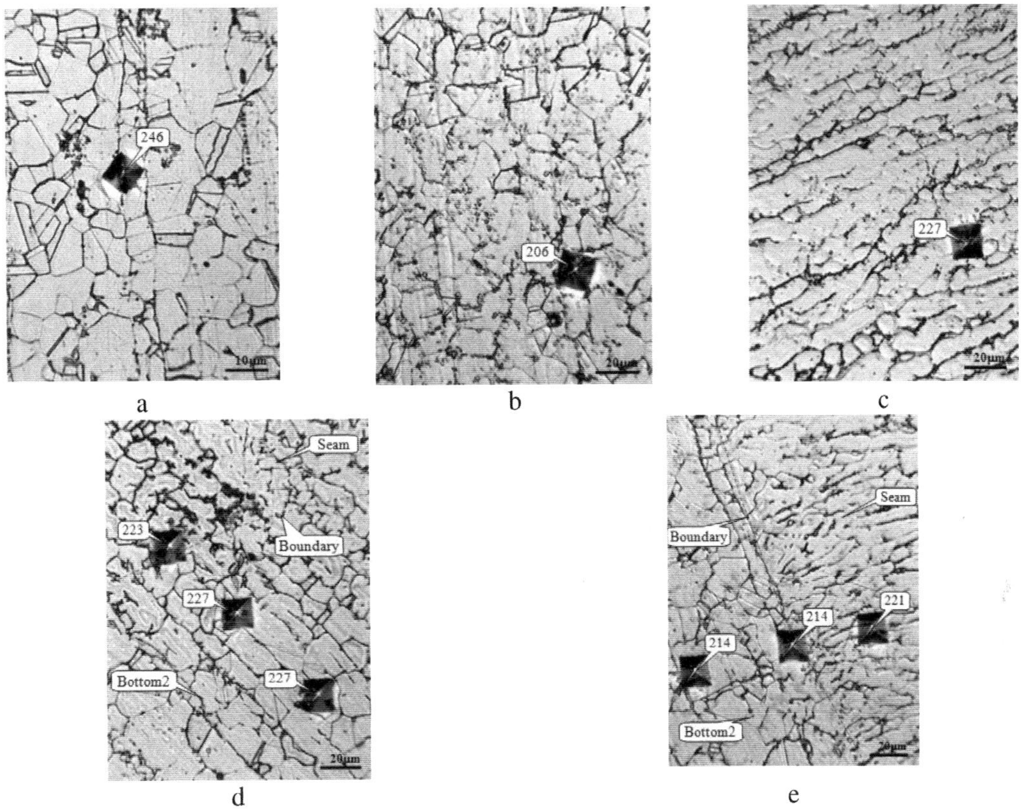

Figure 9. Micro-hardness HV0.05 for Sample 2: a and b base metal zone; c - weld metal zone; d - and e - alloying zone

7. X-ray structural phase analysis

The X-ray structural analysis is one of the most commonly used methods for testing the fine structure of metals. It is based on the ability of X-rays to diffract from crystalline objects that serve as a diffraction grating. For the purpose of this study, an X-ray structural analysis has been made of the welded samples and of some additionally prepared control samples of austenite sheet steel SS316. The purpose was to determine the phase composition in the steel before and after welding with the selected GMAW mode and type of electrode.

From the analysis thus made using the X-ray apparatus model DRON 4, it is clear that the base material and that of the weld seam have phases of the same type, which are mainly austenite and ferrite (Figures 10 and 11).

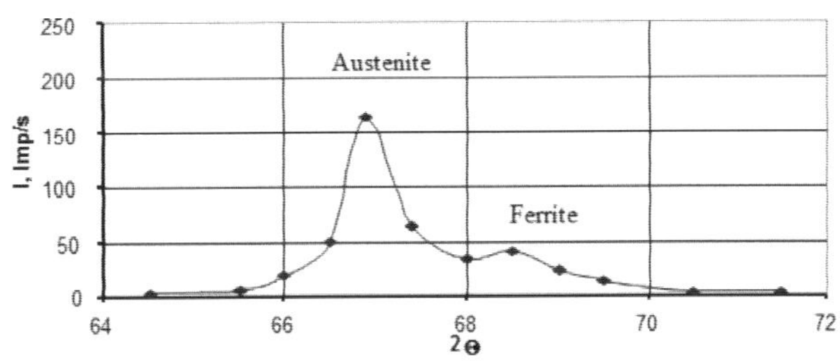

Figure 10. X-ray structural analysis of sheet steel SS316

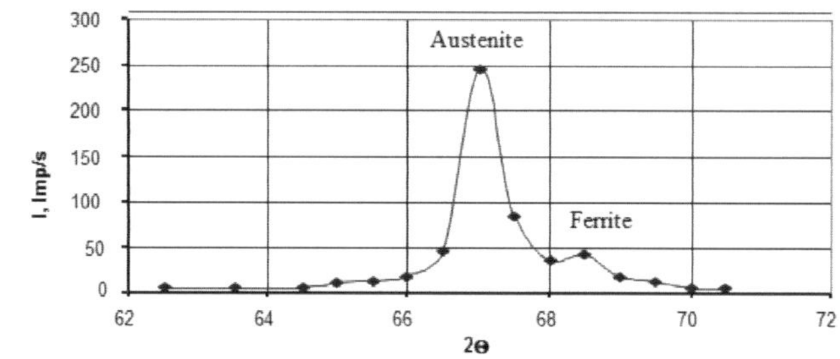

Figure 11. X-ray structural analysis of the welded sample

8. Static tension test of the welding joint

The mechanical properties are of great importance in the production of welding joints. That is why the object of our study was testing the tensile strength of two types of welded samples. For this purpose, 6 laboratory tubes, 3 for each of the two tested sections of the vessel tank were prepared and pre-welded according to the described procedure. They were subjected to static tension with a laboratory tension/compression machine according to BDS EN ISO 7500-1:2016.

The tested samples with their fracture locations indicated are shown in Figure 12. The results of the static tension test of welded joints and sites with fracture zones are shown in Table 3.

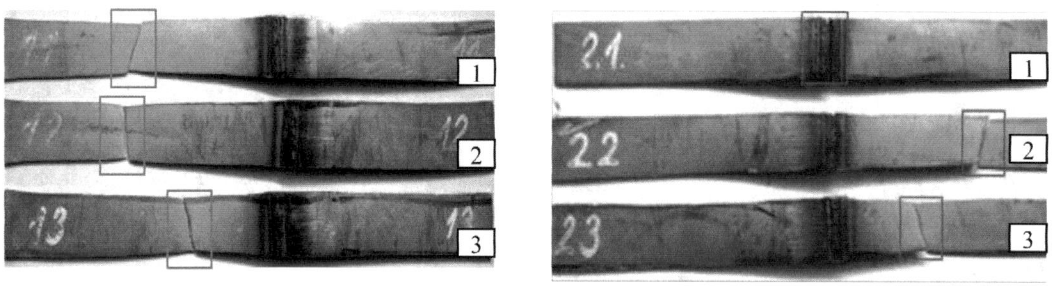

a b

Figure 12. Sample bodies after tension test: a - Sample I; b - Sample II.

Table3. Results from the static tension tests on the welding joints

Tested samples		Measured values of controlled parameters	
№	Sizes* **mm,**	Tensile strength, **Rm, N/mm^2**	Location of destruction
Sample I-1	3.0x19.8	615	In base metal
Sample I-2	3.0x20.5	618	In base metal
Sample I-3	3.0x21.2	603	In base metal
Measured (average value):		**612**	-
Sample II-1	3.0x21.0	489	In weld joint
Sample II-2	3.0x19.3	620	In base metal
Sample II-3	3.0x20.5	618	In base metal
Measured (average value):		**576**	-
Required: for steel 316 accord. To AISI X5CrNiMo17-12-2 accord. To BDS EN 10088-2 material 1.4401.		**500-700**	-

From the tension tests carried out on sample I, it was established that destruction occurs outside the welded joint in the zone of the vessel tank. With Sample II, of the three tension test tubes, only one was destroyed at the welded joint. The cause has been identified and it is due to the displaced and open root of the seam, which indicates the lower strength of the seam material.

9. Conclusion

After the machine GMAW of a vessel tank of austenite steel SS316, the following problems have been diagnosed in the two tested control sections:

✓ For Sample 1:

The root of the seam has penetrated into the wall at the bottom, probably as a result of incorrect positioning of the semi-automatic device, with the profile of the seam not well shaped. As a result of axial and radial stresses, deformation occurred at the point of contact between the cylinder and Bottom 1, which caused cracks to appear symmetrically located at the root (Figure 2 a). In the cross-section of the seam, these cracks also developed in the material of the seam when it was in a liquid and highly plastic state.

There are no gas and slag defects registered in the seam material; it is the evidence of correctly selected modes and the technology of GMAW.

✓ For Sample 2:

The profile of the seam in the tested section is well shaped but again displaced, this time in the direction of the cylinder, due to which part of the contact zone between Bottom 2 and the cylinder is not alloyed; the penetration into the material at Bottom 2 is weak, and into the cylinder it is incomplete. As a result, the root of the seam has a large open surface.

References

[1] Manshadi B D, Mahmudi H, Abedian A, R. Mahmudi R 2007 A novel method for materials selection in mechanical design: Combination of non-linear linearization and a modified digital logic method *Materials and Design* **28** 8–15

[2] Aceves S M, Frias J M, and Villazana O G 2000 Analytical and experimental evaluation of insulated pressure vessels for cryogenic hydrogen storage, *Int. J. Hydrogen Energy* **25** 1075-1085

[3] Wang Y, Liew J Y R and Lee S Ch 2015 Structural performance of water tank under static and dynamic pressure loading *International Journal of Impact Engineering* **85** 110-123

[4] Liu W K 1981 Finite element procedures for fluid-structure interactions and application to liquid storage tanks *Nuclear Engineering and Design* **65** 221-238

[5] Mistríková Z and Jendželovský N 2012 Static analysis of the cylindrical tank resting on various types of subsoil *Journal of Civil Engineering and Management* **18 (5)** 744-751

[6] Ed Obrien R 1998 *Welding Handbook: WeldingProcesses* (AWS)

[7] Bai J W, Liu P P, Zhu Y M et al. 2013 Coherent precipitation of copper in Super 304H austenite steel. *J Mater Sci Eng A* **584** 57–62

[8] Spasova D 2019 Investigation of the abrasion resistance of stainless-steel composites with Al2O3 reinforcement phase produced by using the capillary forming *method IOP Conference Series: Materials Science and Engineering* 012035

[9] Spasova D 2016 Capillary Casting of Iron-Based Alloys *Advances in Materials and Processing Technologies* **2 (03)** 361 – 366

[10] Steel Classification according to EN 10027-2, *Designation systems for steels,* available at: http://www.steelnumber.com/en/steel_composition_eu.php?name_id=81

[11] Product Catalogue ESAB, *Technical Handbook, Stainless Steel Welding,* available at: https://www.esab.bg/

[12] Egerland S 2010 Controlled GMA welding processes prove applicability for high-quality weld overlay *Proceedings of the Welding and Repair Technology for Power Plants: 9th International EPRI Conference* (Fort Myers. Palo Alto: Electric Power Research Institute)

[13] Hardmetals - Metallographic determination of microstructure - Part 1: Photomicrographs and description (ISO 4499-1:2008)

[14] Metallic materials - Vickers hardness test - Part 1: Test method (ISO 6507-1:2018)

IOP Conf. Series: Materials Science and Engineering **843** (2020) 012014 doi:10.1088/1757-899X/843/1/012014

Grinding specifics of plasma coatings melted with high-frequency currents

E A Zverev[1], V Yu Skeeba[1], V Vakhrushev[1] and D V Lobanov[2]

[1] Novosibirsk State Technical University, 20, K. Marksa ave., Novosibirsk, 630073, Russia
[2] Ulianov Chuvash State University, 15, Moscow ave., Cheboksary, 128010, Russia

E-mail: skeeba_vadim@mail.ru

Abstract. This paper presents the research into the finishing treatment of plasma sprayed nickel coatings hardened by high-energy heating with high-frequency currents. We analyzed the coatings surface topography images and profilograms of the after grinding. The coatings were melted at various specific power values and samples traverse speeds relative to the inductor. We also described the different types of surface structural modifications that are formed depending on the melting mode parameters. The research revealed that high-temperature treatment within the efficient range of technological modes results in forming a dense homogeneous structure in the coating with minimal number of pores. The polished surface of such molten coatings has a regular microrelief and a low level of roughness (Ra = 0.305μm). These properties are achieved owing to a significant increase in the structure qualitative properties during the melting process.

1. Introduction

The manufacturing process of applying wear-resistant plasma coatings on the machine parts surface requires finishing machining to ensure the appropriate dimensional accuracy and surface roughness [1-7]. Most often, abrasive grinding is used as a finishing treatment.

The recent improvements in the quality of plasma coatings are due to the combined technologies which involve melting the sprayed layer by a highly concentrated energy source. The high-energy heating with high-frequency currents has been shown to possess a clear advantage at re-processing conductive plasma coatings after the technology options analysis of the main concentrated energy sources (electric arc, electron beam, laser, plasma, and induction) processing methods which can be used for repeated high-energy impact on the structure of plasma coatings [8]. A characteristic feature of this process is that the heating source is volumetric and energy is released at a certain depth of the surface layer, whose value depends on the thermal properties and electrical resistivity of the coating material, as well as on the frequency of the current. In addition, a significant advantage of this method is the technology option to control the depth of induction heating penetration correlating it with the thickness of the applied coating [9].

The research [10, 11] demonstrates that the qualitative properties of the wear-resistant plasma coatings structure increase after heating by high-frequency currents: the number of pores and unmolten powder particles is reduced. Additionally, there is an increase in adhesive strength and wear resistance, which will undoubtedly determine the nature of the finishing machining.

Such defects as pores have significant influence on the micro-relief formation of plasma coatings surface during grinding [12]. When abrasive grains of the grinding wheel penetrate into the coating, the pores become stress concentrators and promote the microcracks development. Notably, in addition to the pores, the surface roughness depends on the quality of the structure. Taking into account the fact that this combined technology significantly affects the sprayed layer properties, the finishing treatment has its own peculiarities whose identification requires research.

The purpose of this work is to study the surface roughness of plasma sprayed coatings made of nickel powder accounting for the specific features of the molten structure after the finishing abrasive grinding.

2. Materials and methods

The studies of surface roughness after finishing machining were carried out on the example of wear-resistant coatings made of self-fluxing nickel powder of the PG-12N-01 brand (fractional particle size is 50 ... 100 μm).

Spraying was performed on samples made of steel 20 with a 40 kW PUN-8 plasma torch. Plasma spraying modes and coating layer thickness were adopted based on previous studies [13, 14].

Coating melting was carried out on an induction unit equipped with the VCHG 6-60/0.44 model generator with an operating current frequency of 440 kHz. Heating was carried out continuously in a sequential manner by the loop inductor equipped with N87 magnet core. Studies on the melting of plasma sprayed nickel coatings were conducted in the specific power range between 3.0×10^8 W/m^2 and 3.2×10^8 W/m^2 with a relative traverse speed between 40 mm/s and 100 mm/s [15, 16]. The depth of energy release ranged between 0.6 and 0.8 mm.

Mechanical finishing was carried out on a 3G71 flat-grinding machine. The workpieces were fixed on the machine table using a magnetic plate. Grinding was performed with a 300 mm diameter abrasive wheel of green silicon carbide with grains grit number 80 (F24). The wheel was of medium hardness (L) with an open structure (8) on a bakelite bond (B). The operating parameters selection was based on the experience of finishing treatment of wear-resistant plasma coatings [12, 17].

The spindle speed was 2250 min-1 (at a cutting speed of 35 m/s). The transverse feed corresponded to 0.3 mm for the double stroke of the table. Taking into account the length of the workpiece and the over travel, the working stroke was equal to 100 mm, and the speed of moving the table in the longitudinal direction was within 11 ... 14 m/min, which corresponded to the frequency of 55...70 double moves per minute. The cutting depth was 0.02 mm, the allowance to be removed was within 0.12 to 0.15 mm, with the thickness of the sprayed coating between 0.65 and 0.70 mm. A water-based solution of NaNO3 (2%) with a flow rate of 2.4 l/min was used as a cooling lubricant. Grinding was performed in several passes and was completed with a sparking-out process.

The topography study and the measurement of the coatings surface roughness after the finishing machining was carried by Zygo New View 7300 research complex.

3. Results and discussion

Figure 1 shows an image of a typical unmolten surface of a plasma sprayed nickel coating obtained using a Carl Zeiss Axio Observer Alm scanning microscope.

The surface of the sprayed coating (Figure 1) has a rough structure with unevenly melted powder particles and pores.

Figure 2 shows an image of the typical topography of the coating surface and the corresponding profilogram after grinding. A characteristic feature of these coatings is an open porosity. The surface profilogram clearly shows a "dip" caused by porosity. On the whole, the processed coatings surface has an irregular microrelief. Undoubtedly, the surface roughness value reflects the number and size of pores. In this case, the surface roughness is Ra = 0.884 μm. In addition to the pores, the roughness depends on such indicators of the structure quality as how melted the powder particles are.

IOP Conf. Series: Materials Science and Engineering **843** (2020) 012014 doi:10.1088/1757-899X/843/1/012014

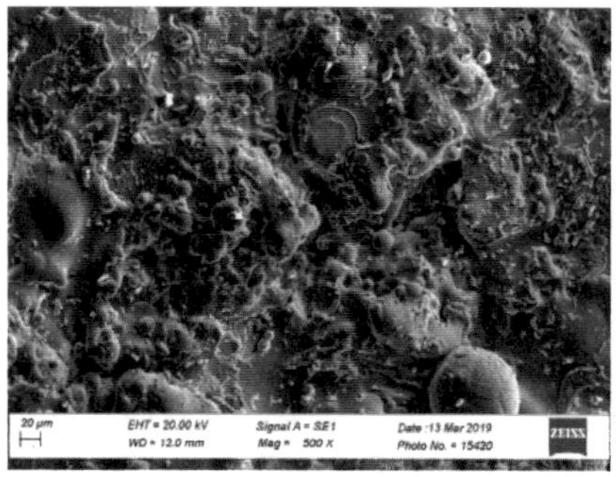

Figure 1. Image of the unmolten surface of the plasma sprayed nickel coating

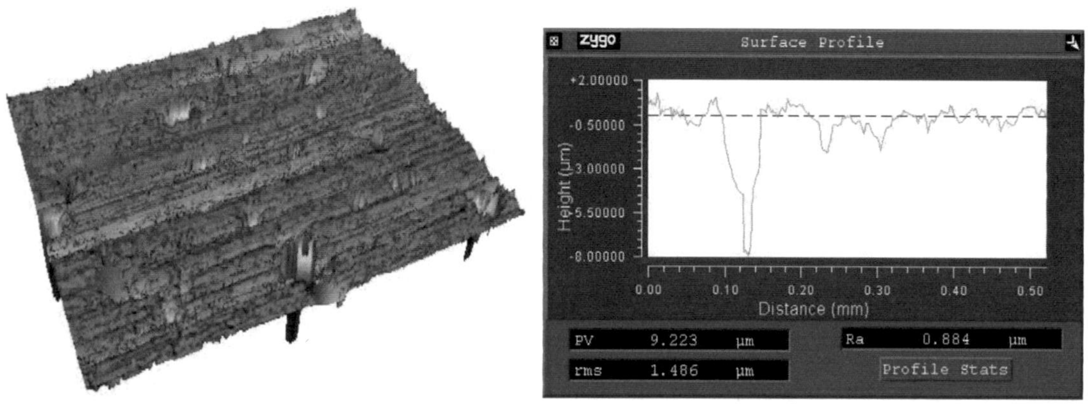

a b

Figure 2. Surface of unmolten coating after grinding: a - topography; b - profilogram

High-energy heating by high-frequency currents forms a dense and uniform structure, with pores and unmolten particles almost completely disappearing, as it is shown in Figure 3.

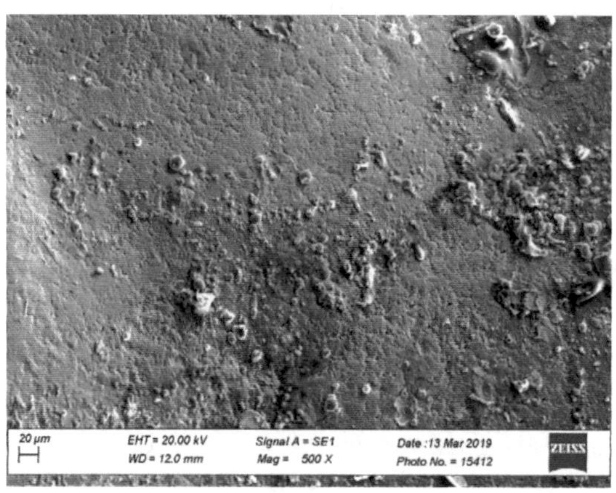

Figure 3. Image of a typical melted surface of the plasma sprayed nickel coating

However, various modifications of the structure can be formed depending on the level of temperature exposure during the melting process. This parameter will undoubtedly affect the surface roughness after finishing treatment.

IOP Conf. Series: Materials Science and Engineering **843** (2020) 012014 doi:10.1088/1757-899X/843/1/012014

The minimum values of the specific power (between 3.0 × 10^8 and 3.1 × 10^8 W/m^2) and the maximum values of the sample traverse speed relative to the inductor (80 to 100 mm/s) result in incomplete melting of the coating whose structure practically does not differ from the initial state. Such coatings' surface after finishing treatment is identical to the polished surface of the unmolten coating.

On the contrary, a structure with an excessive degree of thermal influence is formed at high power (between 3.1 × 10^8 and 3.2 × 10^8 W/m^2) and low sample traverse rates (40 to 60 mm/s). Figure 4 shows the results of the surface study of this structure version after finishing grinding in topography and a profilogram.

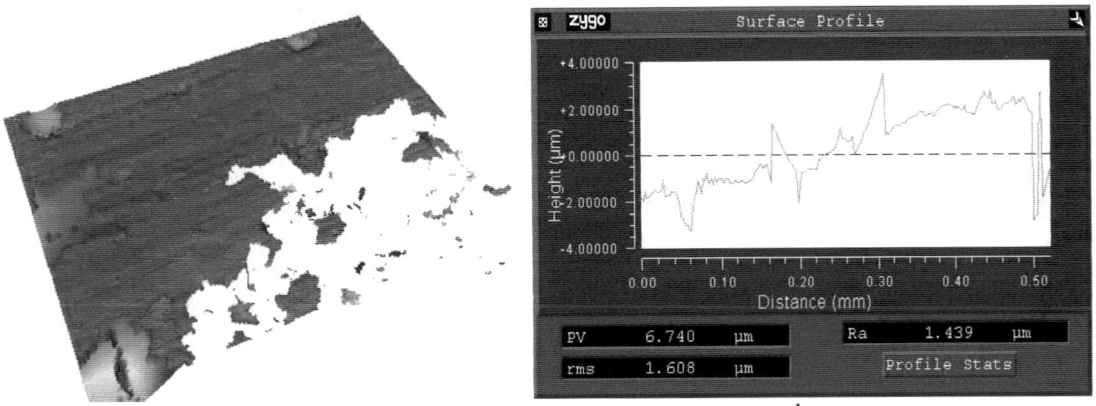

a b

Figure 4. Surface of the coating with excessive melting after grinding: a - topography; b - profilogram

The defective surface after finishing treatment in figure 4 (a) is due to significant overheating during melting, which causes thermal micro-cracks, areas of the material bubbling, burnout of elements and changes in the coating chemical composition. Therefore, the polished surface has a high roughness, with the value of Ra = 1.439 µm.

Given the experience and mathematical simulation techniques of heating by high frequency currents [9, 18], we defined the most efficient modes of subsequent induction high-energy impact: power density range between 3.1·10^8 W/m^2 and 3.2·10^8 W/m^2 at a relative traverse speed between 70 mm/s and 75 mm/s.

A uniformly molten structure with stable quality characteristics is formed when processing coatings in the set operating modes. According to [8, 10, 11], the metallographic analysis of cross-sections at the transition boundary between the coating and the substrate shows a uniform character, there are practically no discontinuity defects, which is clearly reflected in the adhesive strength increase.

Figure 5 shows the corresponding profilogram and topography of the molten coating surface after grinding.

The surface of the melted nickel coating after grinding in this case has a low roughness Ra = 0.305 µm, which corresponds to the level of finishing treatment.

The polished surface of the coating has a regular character. There is no open porosity, which significantly affects the microrelief. The increase in the qualitative characteristics of the coating structure caused by the high-frequency currents melting has a significant impact on the final surface roughness.

IOP Conf. Series: Materials Science and Engineering **843** (2020) 012014 doi:10.1088/1757-899X/843/1/012014

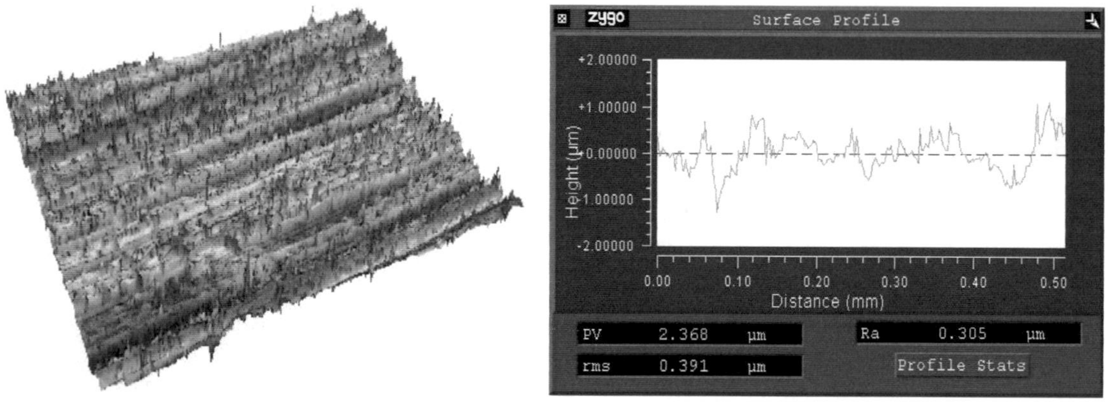

a b

Figure 5. Surface of coating melted at the efficient modes after grinding: a - topography; b - profilogram

4. Conclusion

Studies on the final machining of unmolten plasma sprayed nickel coatings showed that the roughness value after grinding is significantly affected by the number and size of pores, as well as by the presence of unmolten particles.

High-temperature heating of plasma sprayed nickel coatings structure by high-frequency currents within the efficient technological mode allows eliminating these shortcomings and forming a dense structure practically without pores and unmolten particles.

A surface with a regular microrelief and low roughness Ra = 0.305 μm is formed after finishing processing of such molten coatings.

However, it is necessary to take into account that the formation of a structure with insufficient or excessive temperature exposure should be avoided during plasma coatings melting. Overheating is especially undesirable, since it leads to changes in the chemical and phase composition, the appearance of thermal cracks in the coatings and to a rough surface formation after finishing grinding.

5. Acknowledgments

This work was financially supported within the framework of the NSTU Research and Development Thematic Plan (project No. TP-PTM-1_20 project).

References

[1] Lobanov D V et al. 2017 *IOP Conf. Ser.: Earth Environ. Sci.* **87 (8)** 082029
[2] Lobanov D V et al. 2016 *IOP Conf. Ser.: Mater. Sci. Eng.* **142 (1)** 012081
[3] Yanyushkin A S et al. 2015 *IOP Conf. Ser.: Mater. Sci. Eng.* **91 (1)** 012051
[4] Zykova A V et al. 2019 *Materials* **12(6)** 981-990
[5] Liverani E et al. 2016 *Surface and Coatings Technology* **302** 100-106
[6] Aman Y et al. 2012 *Journal of Materials Science* **47 (15)** 5766-5773
[7] Okovity V A, Panteleenko A F 2015 *Obrabotka metallov* **2** 39-45
[8] Chesov Y S et al. 2014 *Obrabotka metallov* **4** 11-18
[9] Skeeba V Y et al. 2016 *IOP Conf. Ser.: Mater. Sci. Eng.* **126 (1)** 012016
[10] Ivancivsky V V et al. 2018 *IOP Conf. Ser.: Earth Environ. Sci.* **194 (4)** 042006
[11] Ivancivsky V V and Skeeba V Yu 2018 *Hybrid metal working equipment. Technological aspects of integrating the operations of surface hardening and abrasive grinding* (Novosibirsk: NSTU Publ.)
[12] Zverev E A et al. 2019 *IOP Conf. Ser.: Earth Environ. Sci.* **378** 012035
[13] Zverev E A et al. 2017 *Key Engineering Materials* **736** 132-137

[14] Zverev E A et al. 2017 *IOP Conf. Ser.: Earth Environ. Sci.* **87** 082061
[15] Chesov Y S and Zverev E A 2015 *In the World of Scientific Discoveries* **3 (1)** 16-26
[16] Chesov Y S and Zverev E A 2015 *In the World of Scientific Discoveries* **2 (62)** 248-266
[17] Zverev E A et al. 2019 *Actual Problems in Machine Building* **6 (1-4)** 174–178
[18] Ivancivsky V V et al. 2018 *IOP Conf. Ser.: Mater. Sci. Eng.* **327 (4)** 042042

IOP Conf. Series: Materials Science and Engineering **843** (2020) 012015 doi:10.1088/1757-899X/843/1/012015

Influence of characteristics of abrasive wheels on surface qualities of high-speed plates HS10-4-3-10 at flat grinding

Nguyen Van Canh[1], **Tran Trung**[1], **Vanthan Nguyen**[1] **and Nguyen Ngoc Hue**[2]

[1] Ngo Quyen University, 229B, Bach Dang st., Binh Duong, 75153, Viet Nam
[2] Thu Dau Mot University, 6, Tran Van On st., Binh Duong, 75151, Viet Nam

E-mail: vancanh.vn@mail.ru

Abstract. The article presents the results of the influence of characteristics abrasive wheel on the surface qualities of high-speed plates HS10-4-3-10 at flat grinding with using statistical method and fuzzy logic simulation. Parameters of surface quality were estimated according by: microrelief R_a, R_{max}, S_m; accuracy form EFE_{max}, EFE_a; microhardness HV, for each parameter is represented by median and quartile latitude. Results of fuzzy logic simulation can use for selection of based elements of the wheel characteristic, to optimize each parameter of surface quality. Optimal abrasive wheel by all parameters in the greatest measure responses predicted as follows: 25AF46K10V5-PO3, 5NQ46I6VS3 and 5SG60K12VXP with the maximum values of the desirability function. Receiving results will be used for optimization of grinding process according by theory of experimental design.

1. Introduction

The grinding process has some features. First of all, the cutting capacities of abrasive tool does not seem possible to characterize the deterministic value, because of the rest of the blade machining processes. As is known, abrasive grains have an arbitrary shape, chaotic arrangement in a bundle, different heights in the radial direction. On this basis, they are divided into cutting, pressing and non-working. Subsequently, the grains differ in the amount of edges, cutting-point angle and radius of there curvature. In the first approximation, the microroughnesses of the part is considered a "mould" of the working surface of the abrasive tool. The high speed and intensity of the heat source cause structural-phase transformations in the surface layer of high-speed cutting plates (HSCPs) [1, 2], leading to a decrease in the resistance of precast cutting tools.

High-speed cutting steel HS10-4-3-10 has an increased tendency to decarburization, reduced toughness, increased wear resistance, reduced grindability. They destine for the manufacture of tools used in the processing of high-strength stainless and heat-resistant steels, alloys and improved alloy steels in conditions of increased heating of the cutting edge: gear-cutting tools, milling and shaped tools, countersinks, taps. Table 1 show chemical composition in % of material weight and steel grades of different standards of HS10-4-3-10 [3].

The problem of grinding process is multi-purpose. It is provided for prediction, increasing of efficiency process, optimization of adaptive control and other methods of the product life of abrasive wheel: manufacturing, working conditions, repair and adjustment processes... For each of them there are special requirements to identify the quality (accuracy form, local of parameters, operational efficiency) of the objects involved [1].

Table 1. Chemical composition of steel HS10-4-3-10 in % of material weight and steel-grade of different standards [3]

Chemical composition of steel HS10-4-3-10 in % of weight by ISO 4957:2004					
C	Cr	W	Mo	V	Co
1.20-1.35	3.8-4.5	9-11	3-4	3.2-3.7	9.5-10.5
Steel grades HS10-4-3-10 according by different standards					
ISO 4957	GOST5950	ASTM A600	JIS 4403	EN 10028-2	SS14
EU	RUSSIA	USA	JAPAN	ENGLISH	SWEDEN
HS10-4-3-10	P9M4K8	T42	SKH57	1.3207	2736

The purpose of current paper is to selecting the optimal abrasive wheel on a surface topography while the surface grinding high-speed cutting steel HS10-4-3-10 using statistical method and fuzzy logic simulation for increasing of the efficiency of grinding process..

2. Materials and methods

The procedure is organically divided into stages: The natural experiment and analysis data of experiment using statistical methods and fuzzy logic simulation.

2.1. Experiment

The experiments were conducted by the periphery of abrasive wheels: abrasive wheels standard porosity - standard porosity, most often the 6th structure; highly porous wheels - 10 - 12th structures. The removal of the operating allowance corresponded to a pendulum scheme without standard grooming at the end of the grinding cycle. Following unchanged conditions were adopted: surface grinding machine 3G71M; the subject of experiment – cylinder high-speed cutting plates HS10-4-3-10 ($HV_{initial}$=7200 mPa) with the size 40x60 mm; abrasive wheels in below part; technological parameters: speed of wheel – v_w=35 m/s, cutting deep t=0.015 mm, the operating allowance z=0.15 mm, longitudinal feed s_l=7 m/min, cross-feed s_c=1 mm/double stroke; cutting fluids – 5% lubrication of emulsion Akvol-6 (TU 0258-024-00148845-98) with speed of flow 7-10 l/min.

The characteristic of wheels area $i=\overline{1;16}$ contain the following information by GOST P52381-2005 (GOST – Russian State Standard), GOST P52587-2006 and catalogs [4, 5]:

- Standard-porous wheels of different manufacturers ($i=\overline{1;5}$): 1–5A46L10VAX, 2–EKE46K3V (corporations Abrasives Corundum, Germany); 3–5NQ46I6VS3 (Norton Vitrium); 4–92A/25AF46L6V20, 5–34AF60K6V5 (Russia).
- High-porous wheels of Norton company ($i=\overline{6;9}$): 6–5SG46K12VXP, 7–5SG46I12VXP, 8–5SG60K12VXP, 9–TGX80I12VCF5 (Altos);
- High-porous wheels of Russian production ($i=\overline{10;16}$): 10–25AF46M12V5-ПО, 11–25AF46M12V5-ПО3, 12–25AF46M10V5-ПО, 13–25AF46M10V5-ПО3, 14–25AF46K10V5-ПО3, 15–25AF60M10V5-ПО, 16–25AF46L10V5-КФ35;

In round Norton Vitrium i=3, as high-porous wheels $i=\overline{6;8}$, number 5 means, betwenness seeds NQ (Norton Quantum) and usual aluminum oxide SG couched in parity. In Russian standard-porous wheel i=4 this condition was realized for the seeds 92A and 25A. In the round Altos (i=9) to seeds TGX were added grain material of sintercorundum SG in parity. Marking of rounds $i=4;5;\overline{10;16}$ are in following link [5], although there is a letter F before the number of grain. To the greatest extent, the designations of characteristics differ for wheels from monocorundum i=1; 2, in which the marking of grains depends on method of making a round: 5A (i=1) is pressurized, EKE (i=2) is cast. The designation of the other constituents of the characteristics of the wheel i = 1 coincides, to the greatest extent with the Russian, except of the structure. In Russian instruments, the numbers of structures are 10; 12 - belong to the highly porous, but the wheel 5A - 10th structure is classified open, and (13 - 20) -the according to the catalog [4] are considered highly porous. In the cast tool i = 2, index "K" indicates the average structure and the number "3" indicates the hardness "soft".

Parameters of the surface quality of the HSCPs were selected: the greatest roughness, specifically – R_{ai}, R_{maxi} and S_{mi}; accuracy form – EFE_{maxi} and EFE_{ai}; microhardness HV_i. Characteristic of parameters and the method of their measurement are given in [2, 6, 7].

2.2. Statistical methods and fuzzy logic simulation

The cutting capacities of abrasive tool are considered variable values. For the interpretation of observations probability-theoretical approaches are used [2, 8-9]. Reduction of the labor intensity of statistical computations is achieved by using of software products, in particular, Statistica 10. In forced of analysis of the experimental data:

$$\left\{y_{ijv}\right\},\ i=\overline{1;16},\ j=\overline{1;6},\ v=\overline{1;30}, \tag{1}$$

where j – number of parameters, v – number of measurement for each parameter. It is possible to lead with the involvement of parametric (average y_{ij}, standard deviations SD_{ij} and ranges $R_{ij}=y_{maxij}-y_{minij}$) and nonparametric (median y_{ij} and quartile latitude $QL_{ij}=|y_{0.75}-y_{0.25}|$) statistical methods. Each of them has "its own field" [8] for effective exertion in technical applications. For the parametric method, it is necessary that all Eq.(1) possess the properties of homogeneity of variance deviation and normality of distributions. Otherwise, the exact criteria of this method lose their reliability and can lead to the adoption of incorrect statistical decisions. In a similar situation, it is more expedient to use ranking statisticians which are not related to any family of distributions and do not use its properties.

The choice of statistical methods is described in [8, 9]. Throughout of this study, we confine ourselves to stating that the procedure for interpreting Eq. (1) has been reduced to one stage as follow: a one-dimensional variance analysis for detecting a significant difference between the levels of measures positions without their nominal search.

The theory of fuzzy logic simulation opens up new possibilities in classification problems, optimal conditions and optimal process of manufacturing which in contrast to traditional statistical methods and allows using the more adaptable methods to represent the initial data of structure [10, 11]. Fuzzy logic simulation is equivalent to theory of fuzzy sets Aij which classes with defuse boundaries represented by sets of ordered pairs, composed of elements y_{ijv} of universals sets $\{y_{ijv}\}$ and the corresponding of membership $\mu_A(y_{ijv})$:

$$A_{ij}=\left\{\left(y_{ijv},\mu_A(y_{ijv})\right)|\ y_{ijv}\in\{y_{ijv}\}\right\}, \tag{2}$$

where $\mu_A(y_{ijv})$ is the characteristic functions that indicate the degree of membership y_{ijv} of fuzzy sets A_{ij}.

The desirability function (d) of fuzzy system was proposed by Harrington which is determined by the expression [10, 11]:

$$d=\frac{\sum_{u=1}^{s} z_u S_u(z_u)}{\sum_{u=1}^{s} S_u(z_u)},\ u=\overline{1;s} \tag{3}$$

where s - is the number of functions belonging to the output term in the base rule; z_u - output value of the system corresponding to the rule having membership function; S_u, $S_u(z_u)$ - the value of the membership function of the term, equal to truth degree of sub-clause of this rule.

The simulation process the fuzzy logic in the Matlab environment, was used a special Fuzzy Logic Toolbox extension package.

3. Results and Discussions

The results of the experiment according by median (med.) and QL_{ijv} after statistical processing (Table 2) are consider input variables for modeling topography of the surface of HSCPs HS10-4-3-10 in the fuzzy logic simulation.

Table 2. Input variables for the simulation of cutting capacities of wheels in the fuzzy logic toolbox

Wheel	Parameter											
	Ra_i, μm		$Rmax_i$, μm		Sm_i, μm		$EFEmax_i$, μm		$EFEa_i$, μm		HV_i, mPa	
	med.	QL	med.	QL	med.	QL	med.	QL	med.	QL	med.	QL
1	0.056	0.015	0.363	0.071	56.22	14.00	16.00	2.75	11.42	2.50	5824.95	734.35
2	0.060	0.012	0.408	0.100	60.19	19.81	15.00	1.00	9.38	1.25	8568.03	1654.71
3	0.049	0.023	0.329	0.153	50.64	9.72	15.00	1.75	11.29	0.83	8895.90	882.68
4	0.065	0.025	0.395	0.109	42.18	13.86	15.00	1.00	11.17	2.88	7190.38	933.58
5	0.057	0.014	0.347	0.088	46.22	13.30	17.00	3.75	12.46	2.69	6863.94	596.74
6	0.047	0.017	0.307	0.129	57.05	17.97	15.00	3.00	11.46	1.90	7747.19	615.91
7	0.050	0.018	0.353	0.138	50.30	18.88	18.00	2.75	12.00	3.54	7918.75	620.63
8	0.048	0.020	0.347	0.126	53.29	32.20	14.00	5.25	8.75	2.13	8033.13	486.09
9	0.055	0.019	0.367	0.111	48.96	21.87	16.00	6.25	10.25	5.88	6002.68	1424.52
10	0.070	0.030	0.463	0.204	79.28	33.17	12.00	2.75	7.75	1.51	9445.00	896.80
11	0.080	0.020	0.510	0.143	50.59	22.88	18.00	4.75	10.42	2.35	7861.56	817.97
12	0.080	0.040	0.550	0.249	67.78	33.21	12.50	3.00	7.75	1.98	10553.73	1294.39
13	0.095	0.133	0.675	0.709	73.15	32.86	16.50	4.75	9.75	3.69	9092.63	811.40
14	0.079	0.021	0.511	0.133	65.12	22.32	12.00	1.00	9.17	1.88	8376.25	714.85
15	0.080	0.030	0.550	0.222	68.53	32.76	14.75	6.88	8.17	4.97	7487.75	501.53
16	0.120	0.150	0.820	0.923	66.22	35.69	18.00	6.00	11.54	3.60	7416.25	776.81

As the table 2 shows, the cutting capacities of wheels are the most correlated with the heights of microroughnesses. In particular, their growth from the median R_{max}0.307(0.32) for the Norton wheel (i=6) to 0.82(1.0) μm for the high-porous wheels 25AF46L10V5-KF35 (i=16) was noted. Here categorical values are placed in parentheses (GOST 2789-73). We have a deterioration of the topography of the surface by the greatest roughness of six categorical values. At the same time, for the analyzed instruments parameter R_a(med.) increased from 0.047(0.05) to 0.12(0.125) μm, respectively for five categorical values. Concerning the average step irregularities S_{mi} in the longitudinal direction, it is worth noting the wheels i=4; 5; 9 that made it possible to obtain them within the limits of categorical value not more 50 μm. At the same time, steps increasing to S_{mi}, i= 10; 13 to categorical value 80 μm, on three values from the regulated series (GOST 2789-73). In the design documentation, the state of the surface of the HSCPs often regulated by parameter R_a. On this parameter, Norton wheels i=3; $\overline{6;8}$ are undisputed leaders, which are providing a micron categorical value 0.05 μm. The choice of the optimum characteristics of the abrasive tool by this parameter R_a allows reducing the number of transitions in the developed grinding operation or increasing it productivity while maintaining the roughness. In similar conditions, cutting capacities of wheels increased the accuracy of the HSCPs form by one qualifier value (GOST 24643-81) and microhardness of their surface increased by 1.6-1.8 times. In both cases, the best results showed wheels i=12; 16. Even, by taking into account medians, it is difficult to choose a wheel that would ensure the best state of the topography of the HSCP surface. If the analysis includes QL, then this problem cannot be realized on basis of statistical methods.

Table 3 presents the results of the first stage of fuzzy logic simulation for each parameter by the wheel; they were estimated by the attributes of medians and QL_{ij}:

- For each parameter, the best result by attributes median and QL_i is predicted wheels: Ra – i=1, 2, 5, 6; Rmax – i=1, 2, 5; Sm – i=3, 4; EFEmax – i= 2, 4; EFEa – i=2, 3 and HV – i=8, 15;
- From table 3 it is established that when grinding HSCPs HS10-4-3-10 recommend using abrasive wheels with characteristics: hardness K, grain size F60, number of porous 12.

Table 3. The results of fuzzy logic on the differential estimation of cutting capacities of wheels

Wheel	Desirability function d_{ij} for parameters					
i	$d_{i1}(R_{a1})$	$d_{i2}(R_{max1})$	$d_{i3}(S_{m2})$	$d_{i4}(EFE_{max})$	$d_{i5}(EFE_a)$	$d_{i6}(HV)$
1	0.848	0.828	0.531	0.602	0.581	0.401
2	0.860	0.793	0.500	0.856	0.773	0.210
3	0.767	0.681	0.687	0.745	0.840	0.513
4	0.724	0.776	0.644	0.856	0.545	0.501
5	0.855	0.796	0.545	0.533	0.532	0.661
6	0.843	0.701	0.501	0.578	0.661	0.653
7	0.829	0.716	0.506	0.557	0.506	0.648
8	0.807	0.733	0.442	0.528	0.629	0.826
9	0.805	0.765	0.513	0.527	0.500	0.202
10	0.673	0.628	0.324	0.602	0.722	0.559
11	0.809	0.722	0.505	0.409	0.599	0.513
12	0.568	0.574	0.407	0.578	0.650	0.746
13	0.332	0.500	0.420	0.526	0.510	0.523
14	0.795	0.790	0.500	0.857	0.666	0.556
15	0.673	0.605	0.423	0.516	0.504	0.806
16	0.121	0.287	0.291	0.497	0.511	0.525

As it can be seen from table 3 for selecting abrasive wheel which ensuring all quality parameters is difficult. Solving this problem needs to do the second stage of fuzzy logic simulation. The value for each variables input to be in the numerical interval [0; 1] with linguistic estimation: [0; 0.5] – Bad; [0; 0.5; 0.5; 0.9] – Middle and [0.5; 0.9] – Good. The desirability function of output (integral estimation) including 4 estimates with numerical intervals and linguistic: [0; 0.37] – Low; [0.37; 0.63] – Normal; [0.63; 0.80] – Good and [0.8; 1.0] – High.

Table 4. The influence of characteristics of wheels on the complex evaluation of surface quality of HSCPs

Wheel (i)	$d_{i\bullet}$	Output (class)
5A46L10VAX (1)	0.57	Normal Cutting Capacities
EKE46K3V (2)	0.68	Good Cutting Capacities
5NQ46I6VS3 (3)	0.72	Good Cutting Capacities
92A/25AF46L6V20 (4)	0.67	Good Cutting Capacities
34AF60K6V5 (5)	0.57	Normal Cutting Capacities l
5SG46K12VXP (6)	0.65	Good Cutting Capacities
5SG46I12VXP (7)	0.58	Normal Cutting Capacities
5SG60K12VXP (8)	0.72	Good Cutting Capacities
TGX80I12VCF5 (9)	0.50	Normal Cutting Capacities
25AF46M12V5-ПО (10)	0.60	Normal Cutting Capacities
25AF46M12V5-ПО3 (11)	0.53	Normal Cutting Capacities
25AF46M10V5-ПО (12)	0.55	Normal Cutting Capacities
25AF46M10V5-ПО3 (13)	0.50	Normal Cutting Capacities
25AF46K10V5-ПО3 (14)	0.74	Good Cutting Capacities
25AF60M10V5-ПО (15)	0.57	Normal Cutting Capacities
25AF46L10V5-КФ35 (16)	0.31	Low Cutting Capacities

Table 4 presents the fuzzy simulation result of the second stage. The last has a simple and well-designed interface that make to choosing optimal abrasive wheel for the optimization of grinding process. Integral assessments of the cutting capacities of wheels (table 4) revealed new leaders in ensuring quality surface of the HSCPs HS10-4-3-10. In particular, the abrasive tools i=14; 8; 3 which includes in the first subclass with the integral evaluation of "good" at the first stage of modeling in the fuzzy logic environment showed themselves on the best side: i=14 – twice in the parameters Rmax, EFEmax; i=8 - once by HV; i=3 - twice by Sm, EFEa, i.e. by one - two parameters from possible variables $j=\overline{1;6}$. The correct choice of abrasive tool characteristics is important, when robust design of grinding operations performed by the best (basic) tool. This approach allows optimizing all objective functions with the greatest efficiency.

Modeling of fuzzy logic by size $d_{i\bullet}$ additionally allows analyzing the role of individual components in the characteristics of wheels:

- It is advisable to grind the HSCPs HS10-4-3-10 with abrasive wheels hardness K (medium soft MS1). In particular, it is advisable to use the wheels 5SG46K12VXP (i=6) $d_{6\bullet} = 0.654$ and 25AF46K10V5-PO3 (i=14) – $d_{14\bullet} = 0.735$, which in comparison with wheels 5SG46I12VXP (I-soft M2) and 25AF46M10V5-PO3 (M-average hardness of C1), allows increasing the desirability function in 1.14 and 1.46 times.
- It is established that with the increasing of grain size from 60 (F60) to 46 (F46) is accompanied by a smaller increase in the desirability function compared to the role of hardness: in 1.1 times for high-porous wheels Norton (i=8; 6); for high-porous wheels from grains 25A (i=15; 12) - only 1.02 times.
- It was found that the wheels of 25AF46M12V5-PO (i=10, $d_{10\bullet} = 0.599$) and 25AF46M12V5-PO3 (i=11, $d_{11\bullet} = 0.534$) with structure 12 allowed increasing the quality of HSCPs HS10-4-3-10 compared to the high-porous wheels of 10th structure: 25AF46M10V5-PO (i=12, $d_{12\bullet} = 0.554$) and 25AF46M10V5-PO3 (i=13, $d_{13\bullet} = 0.502$).
- Wheels with the pore-forming agent PO (i=10; 12) are more effective than the high-porous wheels with PO3 (i=11; 13).

4. Conclusion

The contravention of verification of distribution normality of observations, in connection with foregoing information of surface quality HSCPs, the nonparametric statistic is selected that is characterized median and quartile latitude.

In flat grinding of high-speed cutting plates HS10-4-3-10, the optimal abrasive wheel for each parameter of surface quality was predicted: R_a – i=2; R_{max} – i=1; S_m – i=3; EFE_{max} – i=14, 2, 4; EFE_a – i=3 and HV – i=8.

In these well-tried wheels for grinding steel HS10-4-3-10 recommend the abrasive wheel with characteristics: hardness K, grain size F60, number of porous 12.

Optimal abrasive wheel by all parameter of grinded surface quality HSCPs predicted 25AF46K10V5-PO3, 5NQ46I6VS3 and 5SG60K12VXP with the greatest values of the desirability function.

References

[1] Webster J and Tricard M 2004 *CIRP Annals – Manufacturing Technology* **54 (2)** 597-617
[2] Soler Y I and Nguyen V C 2017 *Journal of Engineering and Technological Science* **49** 291-307
[3] ISO 4957:2004 2004 *Tool Steels* (Geneva: International Organization for Standardization)
[4] *Abrasives for industrial market. Carborundum*, available at: https://www.westtool.com/customer/wetosu/vendor/catalogs/carborundum/industrial_catalog_2014/index.html
[5] *Abrasive Technological Excellence. Norton Saint-Gobain*, available at: https://pdfslide.net/documents/nor ton-catalogus.html
[6] Soler Y I and Kazimirov D Y 2010 *Russian Engineering Research* **30** 251-261

[7] Soler Y I, Prokopeva A V and Kazimirov D Y 2007 *Russian Engineering Research* **27** 769-776
[8] Hollanoler M, Wolfe D A and Chicken E 2013 *Nonparametric Statistical Methods* (New York: Wiley)
[9] Wheeler D J and Chambers D S 2010 *Understanding Statistical Process Control* (Knoxville: SPC Press)
[10] Harington E C 1965 *Industrial Quality Control* **21** 251-261
[11] Trautmann H and Weihs C 2006 *Metrika* **63** 207-213

IOP Conf. Series: Materials Science and Engineering **843** (2020) 012016 doi:10.1088/1757-899X/843/1/012016

Structural and parametric synthesis of screw modules of technological machines

V A Evstratov[1], Yu N Linnik[2], V Yu Linnik[2], V I Grigoryev[1] and V A Suxarnikova[1]

[1] Shakhty Automobile and Road Construction Institute (Branch) of Platov SRSPU(NPI), 1, Lenin Sq., Shakhty, Rostov region, 346500, Russian Federation
[2] Federal State Educational Institution of Higher Education, State University of Management, Moscow, 99 Ryazansky ave.,109542, Russian Federation

E-mail: vae602@yandex.ru

Abstract. The article examines the screw modules of technological machines that feed materials under pressure and move loads in the vertical direction to various working bodies. The main disadvantage of these screw modules is low productivity due to the spiral movement of the material, with a significant deviation from the longitudinal axis of the screw module. The paper considers the possibility of increasing the operational performance of screw modules of technological machines by reducing the rotational component of the material movement. This goal is achieved by installing ribs on the inner surface of the module. The angle of ribs installation is determined by the kinematic and geometric parameters of the conveyor auger shaft and the properties of the supplied and moved loads. The results of theoretical studies show that correctly selected geometrical parameters of the ribs can increase the screw module productivity by 18 – 20%.

1. Introduction.

At present, screw modules are part of various machines, assemblies, and technological complexes, transport, and stationary equipment in the mining industry, the construction industry, and the food industry. Along with traditional screw modules, in recent years, shaftless augers and flexible screw conveyors have become widespread. A wide range of applications in the technological transportation and supply of various materials under pressure is due to such advantages of screw modules as reliability and maintainability, tightness, small dimensions, the possibility of continuous transportation in any direction, the simplicity of laying routes between technological machines, and the ability to move toxic cargo [1].

Flexible screw conveyors do not require additional space for their installation and are used both in designed and existing workshops. The object of research is the structure and parameters of the working bodies of technological and transporting machines, including screw modules, processes occurring in arrays of processed and moving materials, and on the surfaces of their contacts with working bodies. These parameters are determined by the physicomechanical properties of these materials based on modern computer technologies for the selection of technical solutions and design features [2].

IOP Conf. Series: Materials Science and Engineering **843** (2020) 012016 doi:10.1088/1757-899X/843/1/012016

During vertical transportation and feeding under pressure, the material in the screw module moves along a spiral path. Moreover, the angle between the direction of material movement and the longitudinal axis of the screw module is about 800. This method significantly reduces the performance of the module [3, 4].

To a large extent, this is due to the lack of adequate mathematical models that give an accurate quantitative and qualitative assessment of the processes occurring in the material array and on the surfaces of its contact with the working bodies of technological machines. A more accurate quantitative assessment of the process of movement of the material in the screw module gives a mathematical description of the process of movement of the material flow, taking into account the shape of its cross-section [5].

2. The apparatus, operating principles, and basic parameters of screw conveyors

To develop mathematical models of the formation of cargo flows in technological and transporting machines and complexes, it is necessary to establish factors that influence the formation of these cargo flows and justify the dependence of the volume of material at the output on influencing factors [6].

The movement of cargo along a helical line with a small pitch significantly reduces the operational performance of the transportation process. To partially eliminate these shortcomings, design measures are carried out aimed at reducing the coefficient of friction of the load on the surface of the blade and increasing the coefficient of friction of the load on the conveyor body [7]. Screw conveyors have cylindrical bodies with a smooth surface, which weakly prevents the rotational movement of the load inside the conveyor.

In technological machines having screw assemblies that feed material under pressure to forming, grinding, and other working bodies, they have bodies with ribs on the inner surface, which prevent the material from turning with the screw.

Consider the possibility of improving the performance of a vertical screw conveyor by installing ribs on the inner surface of the housing.

3. Disadvantages of screw conveyors and ways to improve them

Technological machines and complexes are designed for use in a particular field of application, and without its specification, the choice of optimal parameters is impossible [8]. When performing a theoretical study of the processes occurring in materials under the action of the working bodies of technological machines, an important step is the identification of the medium and, depending on this, the choice of the design scheme and mathematical model of the stress-strain state of the processed material. Analysis of the movement of cargo in a vertical screw shows that the cargo moves in a spiral and the direction of its movement is about 78 – 82 degrees with the longitudinal axis of the conveyor [9]. This indicator is less than the angle of friction between the metal and most goods transported by screw conveyors. Therefore, if the ribs are arranged in a spiral with a rational angle of cutting the helix, then it can reduce the angle between the direction of movement of the load and the longitudinal axis of the conveyor and thus increase its productivity (Figure 1).

The following forces act on the part of the cargo in the intercostal space (Figure 2). Driving force, from the central part of the load

$$F_{dr} = N_c f_c \cos(\beta - \varphi), \tag{1}$$

where $N_c = m_c R \omega_0^2 \left(\dfrac{\sin\alpha \sin\beta}{\cos(\beta - \alpha)} \right)^2$ is the force acting on the load in the intercostal space;

f_c – coefficient of internal friction of the cargo;

φ – the angle between the ribs and the axis of the conveyor (Figure 2);

m_c – mass of cargo;

R – the radius of the screw;
ω_0 – the angular speed of the screw;
α – the angle of the helix of the screw;
β – the angle between the direction of movement of the cargo and the axis of the conveyor.

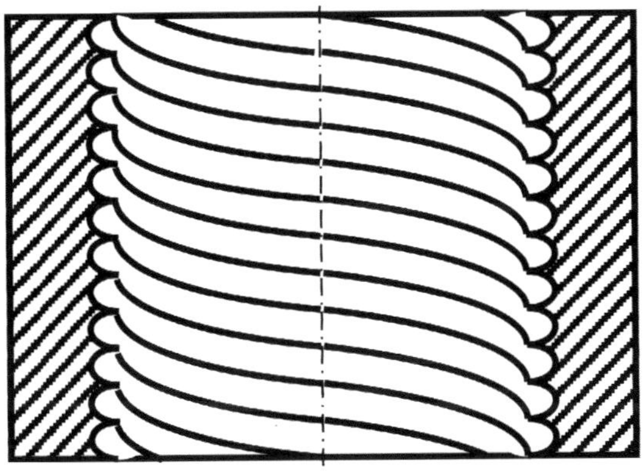

Figure 1. Conveyor tube inner surface profile

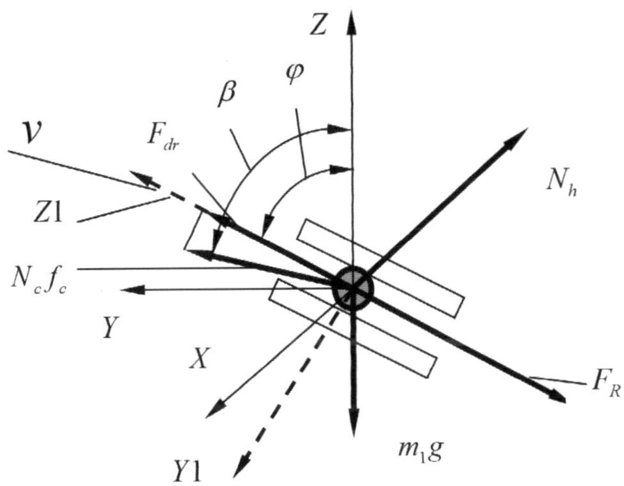

Figure 2. Equilibrium of the part of the flow between the ridge flights.

The force that prevents the movement of cargo between the turns of the ribs

$$F_R = m_1 g f_h \sin\varphi + m_1 g \cos\varphi + N_h f_h . \quad (2)$$

where m_1 is the mass of cargo between the turns of the ribs;
f_h – coefficient of friction of cargo on the body;
$N_h = m_c R\omega_0^2 \left(\dfrac{\sin\alpha \sin\beta}{\cos(\beta-\alpha)}\right)^2 + m_1 R\omega_0^2 \left(\dfrac{\sin\alpha \sin\varphi}{\cos(\varphi-\alpha)}\right)^2$ – the force of the normal pressure of the load moving part and the part located between the turns of the ribs on the conveyor body.

The movement of cargo between the ribs occurs if

$$N_c f_c \cos(\beta-\varphi) > N_h f_h + m_1 g (f_h \sin\varphi + \cos\varphi) . \quad (3)$$

The force that prevents the movement of goods along the ribs

$$F_{Z1} = N_h f_h + m_1 g (f_h \sin\varphi + \cos\varphi). \quad (4)$$

The force that prevents the movement of the load perpendicular to the ribs

$$F_{Y1} = N_c f_c. \quad (5)$$

From dependences (4) and (5), the coefficient of friction of the load along the ribs f_{Z1} and the coefficient of friction across the ribs f_{Y1} are respectively equal:

$$f_{Z1} = f_h + \frac{m_1 g}{N_h}(f_h \sin\varphi + \cos\varphi). \qquad f_{Y1} = f_c. \quad (6)$$

The most significant effect of the ribs will be obtained if condition (3) is satisfied. In this case, the central part of the transported material will move along the surface of the conveyor body under conditions of anisotropic friction.

Consider the movement of cargo in the conveyor when $f_{Z1} < f_{Y1}$. Under these conditions, the direction of the load friction force on the conveyor body does not coincide with the direction of movement of the load (Figure 3). The friction force modulus is

$$F_c = \sqrt{F_n^2 + F_\tau^2}, \quad (7)$$

where F_n – the standard component of the friction force of the load, directed perpendicular to the load velocity vector (Fig. 3); F_τ – the tangential component of the friction force of the cargo, in the direction that coincides with the velocity vector of the cargo.

The friction force of the load on the conveyor body is

$$F_R = N_c f_\beta, \quad (8)$$

where $f_\beta = \sqrt{f_{Z1}^2 \cos^2(\beta-\varphi) + f_{Y1}^2 \sin^2(\beta-\varphi)}$ – the friction coefficient of the load on the inner surface of the conveyor body in the force direction F_R.

The standard component of the friction force of the load on the inner surface of the conveyor housing

$$F_n = N_c f_{(90-\gamma)}, \quad (9)$$

where $f_{(90-\gamma)} = \sqrt{f_{Z1}^2 \sin^2(\gamma-\varphi) + f_{Y1}^2 \cos^2(\gamma-\varphi)}$ – the coefficient of friction of the load on the inner surface of the conveyor body in the direction of the force F_n;

γ – the angle between the direction of movement of the cargo and the longitudinal axis of the conveyor.

The tangential component of the load friction force

$$F_\tau = N_c f_\gamma; \quad (10)$$

where $f_\gamma = \sqrt{f_{Z1}^2 \cos^2(\gamma-\varphi) + f_{Y1}^2 \sin^2(\gamma-\varphi)}$ – the load friction coefficient on the inner surface of the conveyor body in the direction of the force F_τ.

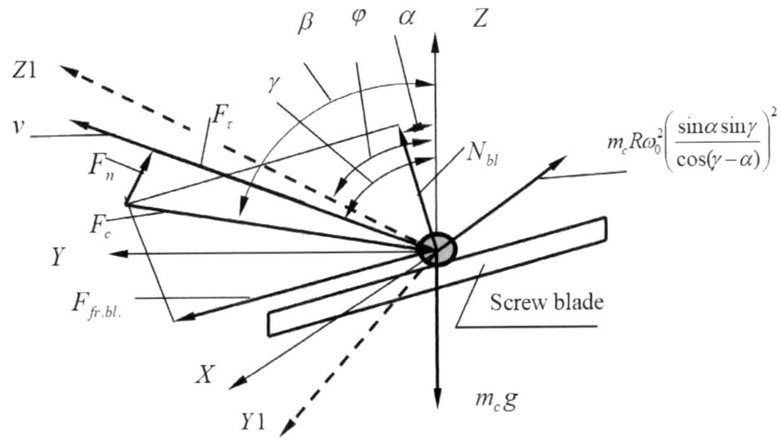

Figure 3. The forces are acting on the load in the conveyor and the direction of its movement when $f_{Z1} < f_{Y1}$.

The equations of motion of a cargo particle [2] do not allow finding the angle γ between the direction of its movement and the axis of the conveyor. To determine the direction of movement of the load on the inner surface of the conveyor housing $f_{Z1} < f_{Y1}$, we use the theorem on the change in kinetic energy.

Since the speed of the load v is maximum in the direction of its movement, the maximum kinetic energy of the load and the sum of the work of all external forces in this direction will also be maximum

$$T = \sum A_i^E = \max. \tag{11}$$

We find the work of all the forces acting on the load in the direction of its velocity vector (Figure 3).

The operation of the normal reaction of a helical blade

$$A_{N_{bl}} = N_{bl} l \cos(\gamma - \alpha), \tag{12}$$

where l – is the magnitude of the movement of the cargo.

The work of the friction force of the load on the blade

$$A_{F_{fr.bl.}} = N_{bl} f_{bl} l \sin(\gamma - \alpha) \cdot \tag{13}$$

The work of the tangential component of the friction force of the load on the inner surface of the conveyor housing

$$A_{F\tau} = -N_c f_\gamma l, \tag{14}$$

where $N_c = m_c R\omega_0^2 \left(\dfrac{\sin\alpha \sin\gamma}{\cos(\gamma-\alpha)}\right)^2$.

Substituting (8, 10, 12 – 14) in (11) we obtain:

$$N_{bl}(\cos(\gamma-\alpha) + f_{bl}\sin(\gamma-\alpha)) - N_c f_\gamma = \max. \tag{15}$$

The angular velocity of the relative movement of the load at $f_{Z1} < f_{Y1}$ is

$$\frac{d\varphi}{dt} = \frac{\omega_0 \cos\alpha \cos\gamma}{\cos(\gamma - \alpha)}. \tag{16}$$

The system of equations [4], which describes the movement of cargo in a vertical screw conveyor at $f_{Z1} < f_{Y1}$, has the following form

$$\left.\begin{aligned} & N_{bl}\cos\alpha - f_{bl}N_{bl}\sin\alpha - f_\beta N_c\cos\beta - m_c g = 0; \\ & f_\beta N_c \sin\beta - f_{bl}N_{bl}\cos\alpha - N_{bl}\sin\alpha = 0; \\ & -N_c + m_c R\omega_0^2\left(\frac{\sin\alpha\sin\gamma}{\cos(\gamma-\alpha)}\right)^2 = 0. \end{aligned}\right\} \tag{17}$$

The joint solution (17) and (15) allows determining the direction of movement of the material in the screw module depending on the angle between the ribs and the axis of the module, the angle of the screw blade, the radius of the screw blade, the angular velocity of the screw, the coefficients of external and internal friction of the material and the radius of the screw the blades.

4. Results

The effect of the ribs on the module case will be more significant; the higher the ratio of the coefficients of internal and external friction of the load. Figure 4 shows the dependence of the angle γ on the ratio of the coefficients of external and internal friction of the load with the following conveyor parameters: radius of the helical blade is $R = 0.2m$, screw blade cutting angle is $\alpha = 20^\circ$; frequency of rotation of a screw blade is $n = 200rpm$; coefficient of internal friction of cargo is $f_c = 1$.

Lines 1 (Figure 4) show the angle of deviation of the direction of the cargo from the vertical under conditions of anisotropic friction at $f_{Z1} < f_{Y1}$ and for various values of the angle of twisting of the ribs φ. Lines 2 reflect the angle of deviation of the direction of movement of the cargo from the vertical with isotropic friction of the cargo against the conveyor body at $f_{Z1} = f_{Y1}$. Lines 3 are the boundaries of the region of propagation of anisotropic friction when the cargo moves along the surface of the conveyor body. The region of anisotropic friction is located on the left side of these lines, and the region of isotropic friction is on the right side, respectively.

The value of the ratio of the coefficients of external and internal friction at points A (Figure 4) at the intersection of curves 1 and 3 is critical. With its further increase, the movement of material in the space between the corrugations ceases, and the direction of movement of the transported material (angle γ) in the corrugated casing becomes the same as in a screw conveyor, the casing of which has a smooth inner surface. On the graph, this is reflected by the transition of the angle γ from point A on curve 1 to point B on curve 2 (angle β).

5. Discussion

The results obtained allow concluding that correctly selected corrugation parameters of the housing can significantly increase the performance of the screw module. The geometrical dimensions and the angle of twisting of the ribs should be determined depending on the particle size distribution and the friction coefficients of the transported goods, in order to achieve the maximum value of the coefficient of material output due to its movement under anisotropic friction.

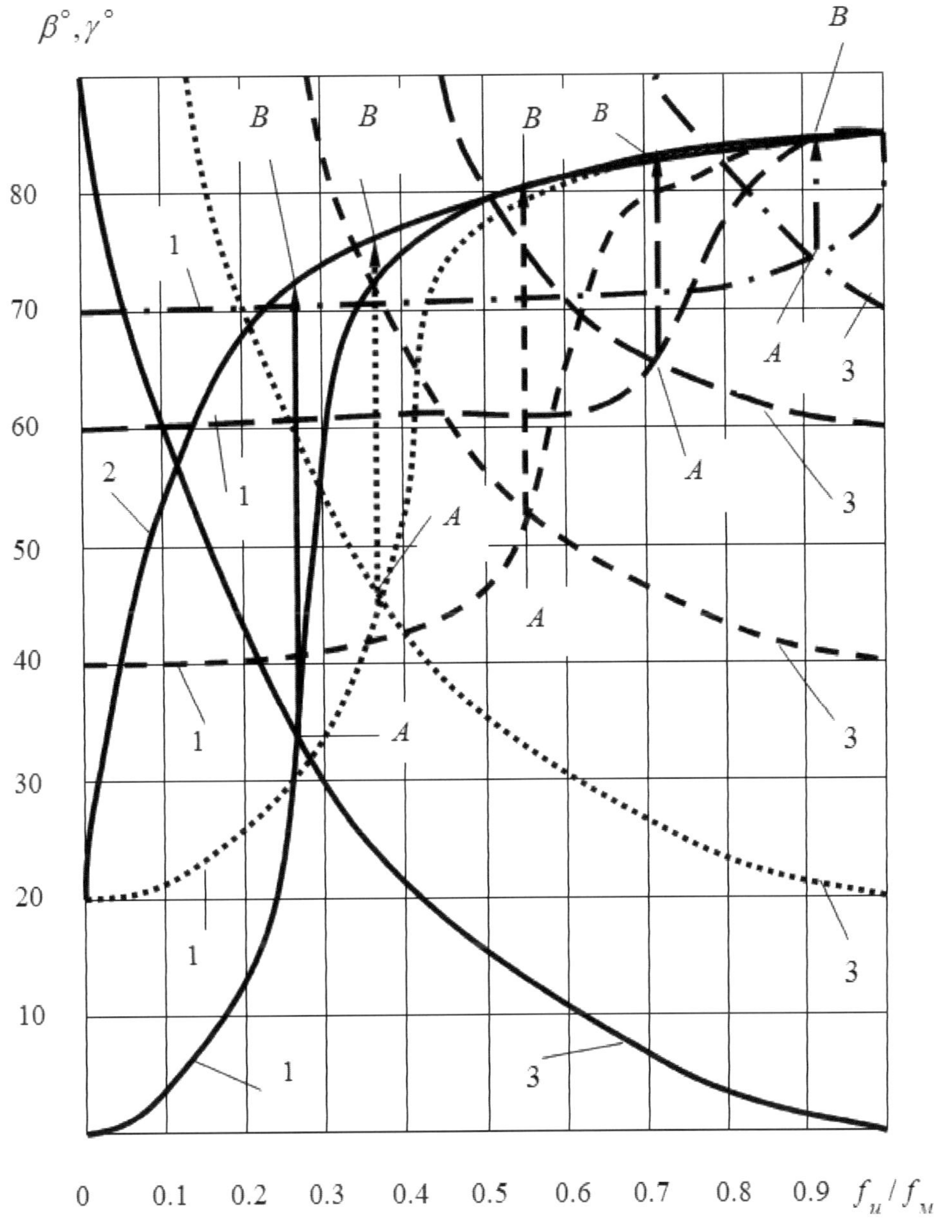

Figure 4. Dependence of the angle between the flow velocity vector and the conveyor axis on the ratio between the flow frictional coefficient and the ridge slop angle φ: ——— $\varphi = 0^\circ$; ▪▪▪▪▪▪▪▪ $\varphi = 20^\circ$; — — $\varphi = 40^\circ$; —— —— $\varphi = 60^\circ$; — ▪ — ▪ $\varphi = 70^\circ$

6. Conclusion

An analysis of the results shows that correctly selected geometrical parameters of the corrugation of the casing can increase the performance of the screw module by 18-20% by reducing the rotational component of the movement of the material.

References

[1] Shemad Zareiforoush, Mohammad Hasan Komarizadeh and Mohammad Reza Alizadeh, 2010 *Journal of Scientific Review* **2 (1)**

[2] Khazanovich G Sh, Lyashenko Yu M, Voronova E Yu and Otrokov A V 2002 *Izvestiya Vysshikh Uchebnykh Zavedenii, Gornyi Zhurnal* **5** 24-27

[3] Philip J Owen and Paul W Cleary 2009 *Proceedings of the Seventh International Conference on CFD in the Minerals and Process Industries* 1- 7

[4] Shemad Zareiforoush, Mohammad Hasan Komarizadeh and Mohammad Reza Alizadeh 2010 (Oriza Sativa L.) *Grains. Nature and Science.* **8(6)**

[5] Rud A V, Evstratova N N, Evstratov V A, Bogdanov D V, Lozovaya S Y, and Lunev A S 2014 *ARPN Journal of Engineering and Applied Sciences* **9** 2376 – 2382

[6] Khazanovich G S, Voronova E Yu and Otrokov A V 2017 *Procedia Engineering* **206** 457-464

[7] Ezzatollah Askari Asli-Ardeh, Ahmad Mohsenimanesh. 2012 *Scientific World Journal* DOI: 10.1100/2012/136218

[8] Voronova E Yu and Khazanovich G Sh 2019 *IOP Conference Series: Materials Science and Engineering* **560 (1)** 012048

[9] Evstratov V A, Rud A V and Belousov K Y 2015 *Procedia Engineering* **129** 397 – 402

IOP Conf. Series: Materials Science and Engineering **843** (2020) 012017 doi:10.1088/1757-899X/843/1/012017

Possibilities for regulation of distribution of oscillation amplitudes points of working bodies of technological machines

A V Eliseev[1], N K Kuznetsov[1], S V Eliseev[2] and Q T Vuong[2]

[1]Irkutsk National Research Technical University, 83, Lermontov St., Irkutsk, 664074, Russia
[2]Irkutsk State Transport University, 15, Chernyshevskiy St., Irkutsk, 664074, Russia

E-mail: eavsh@ya.ru

Abstract. Ensuring the reliability of technological equipment is determined by the dynamic characteristics of operating modes. For objects in the form of extended solids that perform flat or spatial movements, the distribution of oscillation amplitudes at the points of the working body is essential. The problem of adjusting the vibration field of a technological machine is considered. The solution uses methods of structural mathematical modeling, when a mechanical oscillatory system, considered as a computational scheme of a technical object, is compared to a structural mathematical model in the form of the structural diagram of a dynamically equivalent automatic control system. A method for forming certain structures of the vibration field of technological machines is offered. The method is based on the use of additional links implemented by special devices for converting movement. Selection of parameters of a vibrating technological machine by using the effects of changing the reduced mass of devices for converting motion, made in the form of screw non-self-braking mechanisms. The developed mathematical models reflect the vibration field of a homogeneous structure created by "zeroing" angular oscillatory movements. The proposed method for setting the vibration stand parameters is illustrated by computational modeling on the example of a model problem.

1. Introduction

The reliability of technological equipment is determined by keeping the dynamic States of technical objects within certain limits [1-3]. In particular, for objects in the form of extended solids that perform plane or spatial vibrations, it is important to distribute the amplitudes of vibrations at different points of the working body. In many cases, the operating conditions of technical objects are interpreted as problems of adjusting the vibration field of the technological machine. A common way to form the structure of the vibration field is to select the parameters of a mechanical oscillatory system used as a design scheme for a technical object[4,5]. The problems of dynamics of machines and equipment operating under vibration loads require special attention in terms of evaluating the possibilities of improving dynamic states through the introduction of additional links into the structure of mechanical oscillatory systems implemented by various mechanisms and devices for converting motion. A method is proposed for forming certain structures of the vibration field of technological machines based on the ideas of introducing and using additional connections implemented by special devices for converting motion[6-7].

2. Features of technical object

A mechanical oscillating system with a solid body performing flat movements is considered as the design scheme of a vibration stand. A solid body (the working body of the vibrostand) has a mass M and a moment of inertia J (Figure 1). The solid body is based on elastic elements with stiffness k_1 and k_2, which are connected in parallel with additional devices for converting motion (DCM), which have reduced masses L_1 and L_2. Such devices can be implemented on the basis of lever or non-self-locking screw mechanisms. The system has two degrees of freedom and makes small fluctuations relative to the static equilibrium position. Motion is defined in the y_1 and y_2 coordinate system associated with the fixed basis. It is assumed that the external force perturbations Q_1 and Q_2 at the coordinates y_1 and y_2 have the form of common-mode harmonic functions with the same amplitudes, which can be provided by means of centrifugal vibration exciters.

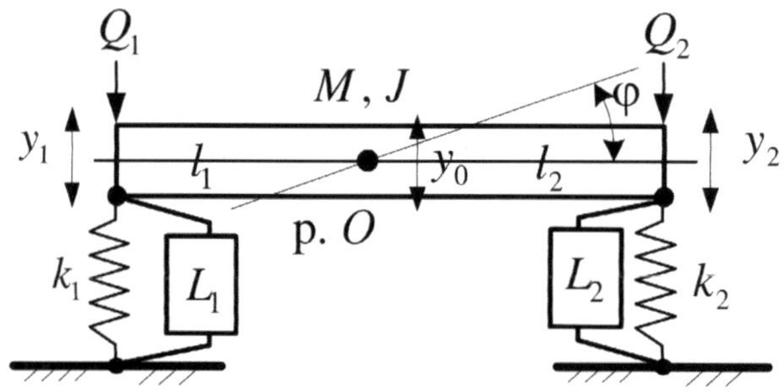

Figure 1. Figure B. 1. Schematic diagram of the vibrostand with a motion conversion device (DCM)

The system of equations of motion can be obtained on the basis of the approaches, using expressions for kinetic and potential energies in the form

$$T = \frac{1}{2}My_0'^2 + \frac{1}{2}J\varphi'^2 + \frac{1}{2}L_1 y_1'^2 + \frac{1}{2}L_2 y_2'^2, \quad (1)$$

$$\Pi = \frac{1}{2}k_1 y_1^2 + \frac{1}{2}k_2 y_2^2. \quad (2)$$

After transformation of Laplace taking into account zero initial conditions the system of equations of motion in operator form takes the form

$$\bar{y}_1[(Ma^2 + Jc^2 + L_1)p^2 + k_1] + \bar{y}_2(Mab - Jc^2)p^2 = \overline{Q}_1, \quad (3)$$

$$\bar{y}_2[(Mb^2 + Jc^2 + L_2)p^2 + k_2] + \bar{y}_1(Mab - Jc^2)p^2 = \overline{Q}_2. \quad (4)$$

Based on the equations (3), (4), a structural mathematical model can be constructed in the form of a block diagram (Figure 2) of an automatic control system.
The transfer functions of the system are obtained on the basis of the block diagram:

$$W_1(p)\Big|_{\overline{Q}_2=0} = \frac{\bar{y}_1}{\overline{Q}_1} = \frac{(Mb^2 + Jc^2 + L_2)p^2 + k_2}{A(p)}, \quad (5) \qquad W_2(p)\Big|_{\overline{Q}_2=0} = \frac{\bar{y}_2}{\overline{Q}_1} = \frac{(Jc^2 - Mab)p^2}{A(p)}, \quad (6)$$

$$W_1'(p)\Big|_{\overline{Q}_1=0} = \frac{\bar{y}_1}{\overline{Q}_2} = \frac{(Jc^2 - Mab)p^2}{A(p)}, \quad (7), \qquad W_2'(p)\Big|_{\overline{Q}_1=0} = \frac{\bar{y}_2}{\overline{Q}_2} = \frac{(Ma^2 + Jc^2 + L_1)p^2 + k_1}{A(p)} \quad (8)$$

$$\underset{\bar{Q}_1=\bar{Q}_2=\bar{Q}}{W_1''(p)}=\frac{\bar{y}_1}{\bar{Q}}=\frac{(Mb^2+Jc^2+L_2)p^2+k_2+(Jc^2-Mab)p^2}{A(p)}, \tag{9}$$

$$\underset{\bar{Q}_1=\bar{Q}_2=\bar{Q}}{W_2''(p)}=\frac{\bar{y}_2}{\bar{Q}}=\frac{(Ma^2+Jc^2+L_1)p^2+k_1+(Jc^2-Mab)p^2}{A(p)}, \tag{10}$$

where

$$A(p)=[(Ma^2+Jc^2)p^2+k_1]\cdot[(Mb^2+Jc^2)p^2+k_2]-[(Jc^2-Mab)p^2]^2 \tag{11}$$

– frequency characteristic equation of the system.

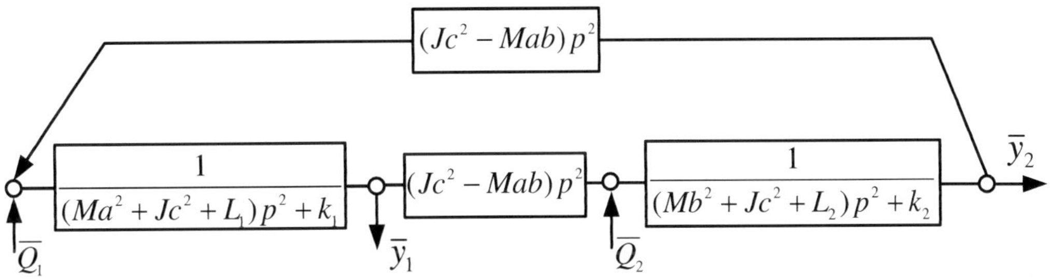

Figure 2. Block diagram of the system according to Figure 1

In the coordinate system $\bar{y}_1$, $\bar{y}_2$, a block diagram has two partial blocks connected by an inertial link. The values of partial frequencies depend on the reduced masses of L_1 and L_2 devices for converting motion (DCM):

$$n_1^2=\frac{k_1}{Ma^2+Jc^2+L_1}, \tag{12}$$

$$n_2^2=\frac{k_2}{Mb^2+Jc^2+L_2}. \tag{13}$$

The system may display modes of dynamic vibration damping, which depends on the conditions of external perturbation. In this case, the frequency of dynamic vibration damping is determined from the condition of "zeroing" the numerators of transfer functions defined by the expressions (5) ÷ (10). The possibilities of distribution of vibration amplitudes along the points of the working body of a vibrating technological machine are determined by the transfer functions of interparticle connections that display the lever properties of dynamic interactions:

$$\underset{\bar{Q}_2=0}{W_{12}(p)}=\frac{\bar{y}_2}{\bar{y}_1}=\frac{(Jc^2-Mab)p^2}{(Mb^2+Jc^2+L_2)p^2+k_2}, \tag{14}$$

$$\underset{\bar{Q}_1=0}{W_{12}'(p)}=\frac{\bar{y}_2}{\bar{y}_1}=\frac{(Ma^2+Jc^2+L_1)p^2+k_1}{(Jc^2-Mab)p^2}, \tag{15}$$

$$\underset{\bar{Q}_1=\bar{Q}_2=\bar{Q}}{W_{12}''(p)}=\frac{\bar{y}_2}{\bar{y}_1}=\frac{(Ma^2+Jc^2+L_1)p^2+k_1+(Jc^2-Mab)p^2}{(Mb^2+Jc^2+L_2)p^2+k_2+(Jc^2-Mab)p^2}. \tag{16}$$

IOP Conf. Series: Materials Science and Engineering **843** (2020) 012017 doi:10.1088/1757-899X/843/1/012017

3. Condition for forming the distribution of oscillation amplitudes

The distribution of the vibration amplitudes at $\frac{\bar{y}_2}{\bar{y}_1} = 1$ corresponds to a situation when a solid body makes translational oscillatory movements vertically in the absence of any oscillatory angular movements. From (16) we get that this is possible if the conditions are met

$$(Ma^2 + Jc^2 + L_1)p^2 + k_1 = (Mb^2 + Jc^2 + L_2)p^2 + k_2, \tag{17}$$

which means that

$$[M(a-b) + L_1 - L_2]p^2 = k_2 - k_1. \tag{18}$$

If it is possible to provide a condition

$$L_2 = M(a-b) + L_1, \tag{19}$$

the ratio of vibration amplitudes $\frac{\bar{y}_2}{\bar{y}_1}$ will be determined by the ratio of stiffness coefficients of elastic elements $\frac{k_1}{k_2}$.

For $k_1 = k_2$ and L_2, defined from (19), it is possible to find conditions for implementing the mode of translational oscillatory vertical movements. If $k_1 \neq k_2$, this mode cannot be implemented. Similarly, we can get a condition when the tuning factor is the value of the reduced mass L_1;

$$L_1 = M(b-a) + L_2. \tag{20}$$

From (19) and (20) it follows, in particular, that the value of the reduced mass used to create a certain dynamic state depends on the difference between the geometric parameters a and b. The smaller the value (a – b), the less the value of the reduced mass is required.
Obtaining a mode close to, it is possible even with one of the devices for converting the movement of L_1 or L_2. Practically, the implementation of the formation of a certain value given weight L_1 or L_2 devices for conversion of movement (DCM) can be carried out by the method described in [7], where the use of the effect of the change is given weight by the application to the nut-flywheel DCM brake torque adjustable, depending on the parameters of the dynamic system state. Figure 3 shows the amplitude-frequency characteristics of interpartial connections. The model problem is used; the data for calculations $a = 0.4$; $b = 0.6$; $c = 1$; $M = 1000$ kg; $J = 250$ kg•m2; $k_2 = 1000$ N/m; $L_2 = 100$ kg; $L_1 = 300$ kg. Graphs of dependencies , that is, the amplitude-frequency characteristics of interpartial connections, are shown in figure 3 (the family of graphs is constructed for $k_1 \to k_2$).

From the analysis of graphs $\frac{\bar{y}_2}{\bar{y}_1}(\omega)$, it follows that in general, when $k_1 \neq k_2$, the amplitude-frequency characteristics of interpartial connections have intersections with the abscissus axis, which determines the modes of dynamic vibration damping ($\bar{y}_2 = 0$). However, the amplitude-frequency characteristics have frequencies at which discontinuities of the second kind are realized, which corresponds to the conditions when . For $\omega = 0$ and $\omega \to \infty$, the dependency graphs $\frac{\bar{y}_2}{\bar{y}_1}(\omega)$ have limit values.

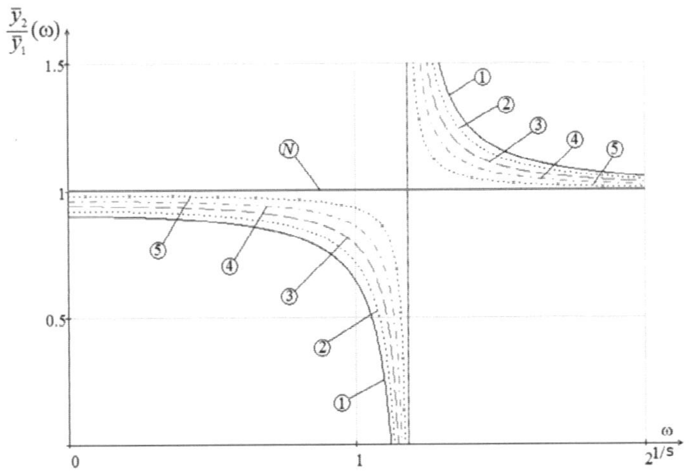

Figure 3. Family of amplitude-frequency characteristics interpartial relations; graphs denoted by (1) represented by curves at k_1 = 900 N/m; graph (2) - k_1 = 920 N/ m; graph (3) - $k1$ = 940 N/ m; graph (4) - k_1 = 960 N/ m; graph (5) - k_1 = 980 N / m; graph marked as (N) corresponds to the operating mode at $\frac{\bar{y}_2}{\bar{y}_1} = 1$

Figure 3 dependency graphs are grouped in such a way that it is possible to observe changes in their relative position when k_1 takes the value k_1 = 900 N/m, 920 N/m, 940 N/m, 960 N/m, 980 N/m approaching a critical case $k_1 = k_2$ = 1000 N/m. In turn, figure 4 shows a family of amplitude-frequency characteristics when k_1 takes the value of 1100 N/m, 1080 nm, 1060 nm, 1040 nm, 1020 N/m, approaching is also the value of k_1 = 1000 N/m and k_2 = 1000 N/m. In the limiting case of the amplitude-frequency response becomes a straight line $\frac{\bar{y}_2}{\bar{y}_1} = 1$, parallel to the x-axis. In this case, the working body of the technological vibrating machine (vibrostand) performs only vertical vibrations in the absence of angular movements. It can be noted that in the zone of sufficient proximity of parameters, when $k_1 \approx k_2$, it is also possible to operate the vibration stand in a certain (or local) frequency range, while $\frac{\bar{y}_2}{\bar{y}_1} \approx 1$.

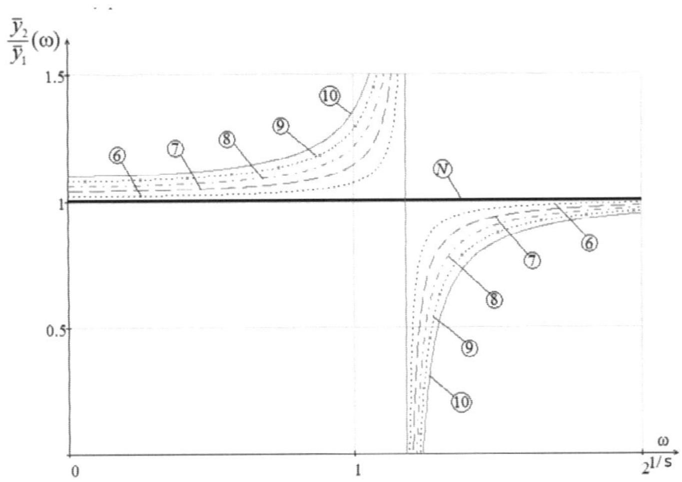

Figure 4. Family of amplitude-frequency characteristics interpartial relations; graphs denoted by (6) represented by curves at - k_1 = 1020 N / m; graph (7) - k_1 = 1040 N/ m; graph (8) - k_1 = 1060 N/ m; graph (9) - k_1 = 1080 N/ m; graph (10) - k_1 = 1100 N / m; graph marked as (N) corresponds to the operating mode at $\frac{\bar{y}_2}{\bar{y}_1} = 1$

Figure 5 shows a family of amplitude -frequency characteristics of interpartial connections with a more detailed consideration of the convergence of coefficients k_1 and k_2 (in particular, the values of parameters k_1 = 995 N/m, 999 N/m, 1001 N/m,1005 N / m).

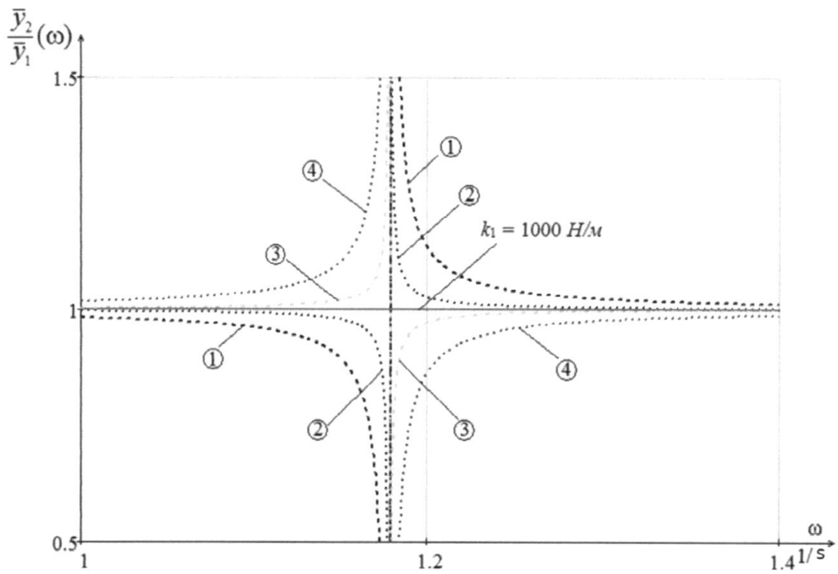

Figure 5. Family of amplitude-frequency characteristics between partial connections; graphs (1) are curves for $k_1 = 995$ N / m; graphs (2) - $k_1 = 999$ N/ m; graphs (3) - $k_1 = 1001$ N/ m; graphs (4) - $k_1 = 1005$ N / m; graphs with the value $k1 = 1000$ N / m corresponds to the operating mode when $\frac{\bar{y}_2}{\bar{y}_1} = 1$

4. Discussion

Comparative analysis shows that with the symmetry of elastic characteristics, that is, at k_1, close enough to the value of k_2, it is possible to adjust the vibrostand by introducing an additional connection that creates compensating effects of changes in the reduced masses created by devices for converting motion, which ensure equality of coefficients at frequency squares in expressions for transfer functions of interpartial connections. The analysis also shows that if the stiffness of the elastic elements k_1 and k_2 significantly mismatch, their preliminary adjustment of the elastic system of the vibration stand is required, which can be implemented when preparing the vibration processing machine for operation.

The implementation of vibration technological processes requires an appropriate preliminary assessment of the dynamic capabilities of vibration stands in terms of creating certain conditions for interaction between the processing working environment and the processed product. The author proposes a method for constructing mathematical models that reflect the features of the formation of a working body of a technological machine of a vibration field of a uniform structure, which is achieved by «zeroing» angular oscillatory movements.

Mathematical models can be constructed based on the use of transfer functions of interpartial relations. The author offers a method for selecting the parameters of a vibrating technological machine by using the effects of changing the reduced mass of devices for converting motion, made in the form of screw non-self-braking mechanisms.

5. Conclusion

The proposed method for setting the parameters of the vibrostand is illustrated by computational modeling on the example of a model problem. A comparative analysis of the possibilities of changing dynamic properties when configuring the working bodies of vibration stands allows using the possibility of adjusting the vibration field within a fairly wide range, which implies the possibility of working not only with a coefficient of connectivity of vibration amplitudes equal to one, over the entire range of the operating mode of the vibration stand, but also in local zones of excitation frequencies, when the coefficient of connectivity of vibration amplitudes fluctuates within a small range relative to one.

Acknowledgments
The reporting study was implemented with the financial support of the Ministry of science and higher education of the Russian Federation in the framework of project No. 0667-2020-0037 "Development of scientific foundations of effective processes obtaining valuable elements, substances and materials from waste and man-made and natural multicomponent mixtures, including methods of design and technological support".

References

[1] Karnovsky I A and Lebed E 2016 *Theory of vibration protection* (Switzerland: Springer)
[2] de Silva C W 2000 *Vibration. Fundamentals and Practice* (Boca Raton, London, New York, Washington, DC: CRC Press)
[3] Harris S M and Piersol A G 2002 *Shock and Vibration Handbook* (New York: McGraw - Hill Book Co)
[4] Kopylov Yu R 2011 Dynamics of the vibration hardening processes: monograph (Voronezh: IPC “Scientific book”)
[5] Panovko G Ya 2006 *Dynamics of the vibration technological processes* (Moscow; Izhevsk: Research Center for Regular and Chaotic Dynamics, Institute of Computer Technologies)
[6] Eliseev A V, Kuznetsov N K and Moskovskikh A O 2019 *The Dynamics of Machines. System conception, structural schemes and connections of elements* (Moscow: Innovative mechanical engineering)
[7] Eliseev S V and Eliseev A V 2020 *Theory of Oscillations. Structural Mathematical Modeling in Problems of Dynamics of Technical Objects* (Springer International Publishing)

Innovations in engineering: analysis of the increase effect in net present value

Yu Kirillov, E Dragunova, A Kravchenko and A Dorofeeva

Novosibirsk State Technical University, 630092, Novosibirsk, K. Marx Ave., 20, Russia

Abstract. The article discusses the construction of mathematical models for assessing economic conditions under which a positive effect of using loan capital in the implementation of innovative and investment projects in engineering is possible, leading to an increase in net present value (NPV). Based on the constructed models, the authors obtained economic and mathematical relations (linking the parameters of innovation and investment projects: investor profitability standard, loan capital rate, time it was entered into the financial and temporal scheme of the project, the time it began to pay off financial obligations and its duration), which determine the increase effect of NPV due to the corresponding structure of investment capital. In the form of functions of the project parameters, analytical expressions are obtained to determine the critical value of the loan rate, the time of its input, as well as the rate of return on the investor and at which the effect of an increase in the net present value of the project becomes possible. The mathematical relations and the economic conclusions drawn on their basis are backed by practical results, given in a graphic form.

1. Introduction

Engineering is a strategic industry, without which a stable, stable, dynamically developing economy is impossible. However, the solution of the task set by the President of Russia to create a "smart" economy is constrained by the fact that modern engineering remains largely in the position of the 4th technological structure, while advanced countries are already actively mastering level 6. Such a technological breakthrough can be realized with the help of appropriate innovative projects, the content of which fully complies with the strategic program [1] to ensure national economic security and strengthen the competitiveness of our country in foreign markets.

From the point of view of achieving the final result, innovative activity can be considered as a process of implementing an investment project [2, 3], the peculiarity of which is the presence of a R&D stage in the investment part of the project and an increased level of risk during its implementation. In Russia, domestic R&D expenditures do not exceed 1.1% of GDP, and among the BRICS countries, Russia is significantly inferior to China (2.1% of GDP) [4], at that time, as the share of R&D expenses in the total revenue of companies from the USA, Germany, Japan and Switzerland increased in 2017-2019 by 4.5% [5]. The data presented indicate that the financing of investment projects related to the development and implementation of innovative technologies needs to be increased. That is why in modern conditions, along with the importance of implementing innovative investment projects, the solution to the question of assessing the effectiveness of such projects becomes equally relevant.

In modern approaches, the effectiveness of innovations is assessed, first of all, by the indicator of net present value, however, the standard algorithm for its calculation does not take into account the structure of the investment capital of the project. In real innovative projects, very often, with a lack of equity, loan capital is attracted, which, on the one hand, makes it possible to realize the investment part of the project, but, on the other hand, reduces its profitable payments by paying off financial obligations. That is why the assessment of the impact, first of all, of loan capital, taking into account its cost and the time it takes to enter the financial and temporal scheme of the project, becomes especially relevant when solving the issue of assessing the effectiveness of innovations.

2. Problem statement

It is necessary to assess the impact that the use of loan capital (LC), as well as equity capital (EC), makes on the indicator of the net present value (NPV) of the innovation investment project (IIP).Obviously, a quantitative assessment of such an impact will be determined by the financial-time pattern of the implementation of IIP. From the point of view of the structure of investment capital, it is considered economically feasible to use the phased financing method [2, 3], the essence of which is that the necessary finances are allocated only at the beginning of the next stage of the project so that by the beginning of the next one decide on whether to continue the project or close it. This allows reducing the risk level of a potential investor in the implementation of IIP to a certain extent.

The financial-time diagram of the fundamental features of the implementation of IIP using the method of phased financing is shown in Figure 1.The specifics of the implementation of specific capital-intensive projects will differ from the above diagram in detail: the amount of capital, the duration of individual stages, but the general structure of capital and the dynamics of its use will correspond to the diagram in Figure 1.

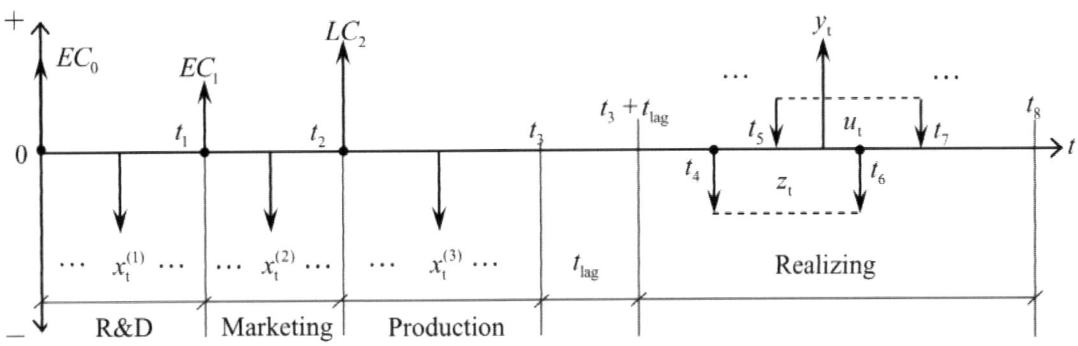

Figure 1. Financial - time diagram for the implementation of IIP

At the beginning of the 1st stage of R&D ($t=0$), financing $x_t^{(1)}$ ($t_1=0,1,...,t_1$) is carried out at the expense of the investor's equity EC_0; at the beginning of the 2nd stage ($t=t_1$) expenses for marketing research $x_t^{(2)}$ ($t_2=t_1+1,...,t_2$) - due to EC_1. If at the 3rd stage ($t_3=t_2+1,...,t_3$) - production of an innovative product - own funds are not enough, loan capital LC_2 is introduced at the moment $t=t_2$, for example, in the form of a loan at a rate of r % per annum.

At the moment $t=t_3$, the investment part of the project ends and with some lag t_{lag}, which corresponds to the main features of the IIP, its revenue part begins. From income from the sale of an innovative product y_t ($t=t_3+t_л,...,t_8$), payments are made u_t ($t=t_5,...,t_7$) to owners, as well as repayment of a loan z_t ($t=t_4,...,t_6$).

To solve the problem of determining the influence of LC and EC on the NPV indicator, it is necessary to perform the following economic and mathematical calculations:

1) the definition of the analytical expression of the NPV indicator as a function of LC, EC, the rate of return on the investor, the cost of credit and time parameters of the project;

2) the definition of analytical expressions of economic conditions for the rate of return and loan rate, in which the effect of the increase in NPV will be positive.

3. Theoretical analysis

In accordance with the rules of financial mathematics, based on a systematic approach to the analysis of investment processes [6], the net present value of an innovative project, the financial-time diagram of which is presented in Figure 1 will be determined by the expression:

$$NPV\ x_t, y_t = \left[P_0\ y_t + P_0\ EC_1 + P_0\ LC_2\right] - \left[P_0\ x_t + P_0\ z_t + P_0\ u_t\right], \quad (1)$$

where $P_0(y_t) = \sum_{t_3+t_{lag}}^{t_8} \frac{y_t}{(1+i_0)^t}$ and $P_0(x_t) = \sum_{0}^{t_1} \frac{x_t^{(1)}}{(1+i_0)^t} + \sum_{t_1+1}^{t_2} \frac{x_t^{(2)}}{(1+i_0)^t} + \sum_{t_2+1}^{t_3} \frac{x_t^{(3)}}{(1+i_0)^t}$ are the values of its inflows x_t and outflows y_t discounted to the beginning of the project, respectively; $P_0(EC_1) = \frac{EC_1}{(1+i_0)^{t_1}}$ – discounted value of equity EC_1 introduced at the time t_1; $P_0(3K_2) = \frac{3K_2}{(1+i_0)^{t_2}}$ – discounted value of the loan capital LC_2 introduced at the time t_2; $P_0(z_t) = \sum_{t_4}^{t_6} \frac{z_t}{(1+i_0)^t}$; $P_0(u_t) = \sum_{t_5}^{t_7} \frac{u_t}{(1+i_0)^t}$ – discounted value of repayment payments z_t and payments u_t to owners and the respective; i_0 – investor profitability ratio taking into account innovative risk. The size of payments u_t is determined by the conditions and the cost of formation of IIP's own capital, and the size of repayment payments z_t is determined by the conditions of full repayment of the loan LC_2 by the time $t = t_6$. If you write (1) as follows:

$$NPV\ x_t, y_t = \left[P_0\ y_t - P_0\ x_t\right] + \left[P_0\ EC_1 - P_0\ u_t\right] + \left[P_0\ LC_2 - P_0\ z_t\right],$$

it $P_0\ y_t - P_0\ x_t = NPV_0$ can be considered as the basic net present value, determined on the basis of the values of inflows and outflows in the business plan of the project; $P_0(EC_1) - P_0(u_t) = \Delta NPV_t(EC_1)$ – as an increase in net present value due to the introduction at the time t_1 of equity EC_1 (taking into account its cost); $P_0\ LC_2 - P_0\ z_t = NPV_t(LC_2)$ – as an increase in net present value due to the introduction at the time t_2 of loan capital LC_2 (taking into account its value). Then

$$\Delta NPV_t = \Delta NPV_t(EC_1) + \Delta NPV_t\ LC_2 = \left[P_0\ EC_1 - P_0\ u_t\right] + \left[P_0\ LC_2 - P_0\ z_t\right] \quad (2)$$

there is a general increase in net present value of IIP due to the use of both own and loan capital. From (2) it obviously follows that the increase effect will be positive if the following conditions are met:

1) $P_0\ EC_1 > P_0\ u_t$ and $P_0\ LC_2 > P_0\ z_t$;

2) $P_0\ EC_1 > P_0\ u_t$ and $P_0\ LC_2 < P_0\ z_t$, but $P_0\ EC_1 - P_0\ u_t > P_0\ LC_2 - P_0\ z_t$;

3) $P_0\ EC_1 < P_0\ u_t$ and $P_0\ LC_2 > P_0\ z_t$, but $P_0\ EC_1 - P_0\ u_t < P_0\ LC_2 - P_0\ z_t$.

In practice, the repayment flow is usually annuity LC_2, and for the vast majority of cases, it is in the form of annuities with annual interest accrual at the rate r and annual payments of $z_t = \text{const}$, which according to the rules of financial mathematics [7] for the diagram in Figure 1

$z_t = \dfrac{LC_2 \cdot r \cdot (1+r)^{(t_4-t_2)}}{1-(1+r)^{-(t_6-t_4)}}$. Then, by the same rules, we get:

$$P_0(z_t) = z_t \cdot \frac{1-(1+i_0)^{-(t_6-t_4)}}{i_0} \cdot \frac{1}{(1+i_0)^{t_4}} = \frac{LC_2 \cdot r \cdot (1+r)^{(t_4-t_2)}}{1-(1+r)^{-(t_6-t_4)}} \cdot \frac{1-(1+i_0)^{-(t_6-t_4)}}{i_0} \cdot \frac{1}{(1+i_0)^{t_4}}$$

And the equitation $P_0(LC_2) - P_0(z_t) > 0$ is written in the form:

$$\frac{LC_2}{(1+i_0)^{t_2}} - \frac{LC_2 \cdot r \cdot (1+r)^{(t_4-t_2)}}{1-(1+r)^{-(t_6-t_4)}} \cdot \frac{1-(1+i_0)^{-(t_6-t_4)}}{i_0} \cdot \frac{1}{(1+i_0)^{t_4}} > 0. \quad (3)$$

Algebraic transformations (3) we obtain the condition

$$\frac{LC_2}{(1+i_0)^{t_2}} \cdot \left\{ \frac{i_0 \cdot (1+i_0)^{t_4-t_2} \cdot \left[1-(1+r)^{-(t_6-t_4)}\right] - r \cdot (1+r)^{(t_4-t_2)} \cdot \left[1-(1+i_0)^{-(t_6-t_4)}\right]}{\left[1-(1+r)^{-(t_6-t_4)}\right] \cdot i_0 \cdot (1+i_0)^{t_4-t_2}} \right\} > 0,$$

which will be executed if

$$\delta = i_0 \cdot (1+i_0)^{t_4-t_2} \cdot \left[1-(1+r)^{-(t_6-t_4)}\right] - r \cdot (1+r)^{(t_4-t_2)} \cdot \left[1-(1+i_0)^{-(t_6-t_4)}\right] > 0.$$

By writing the last expression as

$$\delta = \underbrace{\left[i_0 \cdot (1+i_0)^{\Delta t_1} - r \cdot (1+r)^{\Delta t_1}\right]}_{\delta_1} + \underbrace{\left[r \cdot \frac{(1+r)^{\Delta t_1}}{(1+i_0)^{\Delta t_2}} - i_0 \frac{(1+i_0)^{\Delta t_1}}{(1+r)^{\Delta t_2}}\right]}_{\delta_2} > 0 \quad (4)$$

when $\Delta t_1 = t_4 - t_2$ and $\Delta t_2 = t_6 - t_4$, we can consider various options for determining the effect of increase $\Delta NPV_t(LC_2)$, depending on the ratio i_0 of the rate of return of the investor and the loan rate *r*:

1) If $\boldsymbol{i_0 = r}$, then, as follows from (4) $\delta = 0$ and, moreover, as in the case $\Delta t_1 = \Delta t_2$, so in the case $\Delta t_1 > \Delta t_2$ and in the case $\Delta t_1 < \Delta t_2$. Then

$$\Delta NPV_t(LC_2) = 0; \quad (5)$$

2) If $\boldsymbol{i_0 > r}$, then, as follows from (4) $\delta_1 > 0$ and $\delta_2 < 0$. Therefore, under the condition

$$i_0 \cdot (1+i_0)^{\Delta t_1} - r \cdot (1+r)^{\Delta t_1} > \left| r \cdot \frac{(1+r)^{\Delta t_1}}{(1+i_0)^{\Delta t_2}} - i_0 \frac{(1+i_0)^{\Delta t_1}}{(1+r)^{\Delta t_2}} \right|, \quad (6)$$

$\delta > 0$, moreover, as in the case $\Delta t_1 = \Delta t_2$, so in the case $\Delta t_1 > \Delta t_2$ and in the case $\Delta t_1 < \Delta t_2$;

3) If $\boldsymbol{i_0 < r}$, then, as follows from (4) $\delta_1 < 0$ and $\delta_2 > 0$. Therefore, under the condition

$$r \cdot \frac{(1+r)^{\Delta t_1}}{(1+i_0)^{\Delta t_2}} - i_0 \cdot \frac{(1+i_0)^{\Delta t_1}}{(1+r)^{\Delta t_2}} > \left| i_0 \cdot (1+i_0)^{\Delta t_1} - r \cdot (1+r)^{\Delta t_1} \right|, \quad (7)$$

$\delta > 0$, moreover, as in the case $\Delta t_1 = \Delta t_2$, so in the case $\Delta t_1 > \Delta t_2$ and in the case $\Delta t_1 < \Delta t_2$.

In the general case, the analytical formula for determining the increase effect in the net present value of an innovative project will depend on the type of payment flows z_t of loan LC_2 repayments , as well as payments u_t to owners EC_0 and EC_1:

$$\Delta NPV_{\text{t}} = \left[\frac{EC_1}{(1+i_0)^{t_1}} - \sum_{t_5}^{t_7}\frac{u_{\text{t}}}{(1+i_0)^t}\right] + \left[\frac{LC_2}{(1+i_0)^{t_2}} - \sum_{t_4}^{t_6}\frac{z_{\text{t}}}{(1+i_0)^t}\right]. \quad (8)$$

If the stream u_{t} also represents the annual annuity rent, then, based on (3) and (8), we obtain

$$\Delta NPV_{\text{t}} = EC_1 - u_{\text{t}}\cdot\frac{1-(1+i_0)^{-(t_7-t_5)}}{i_0\cdot(1+i_0)^{t_5}} + \frac{LC_2}{(1+i_0)^{t_2}} - \frac{LC_2\cdot r\cdot(1+r)^{(t_4-t_2)}}{[1-(1+r)^{-(t_6-t_4)}]}\cdot\frac{\left[1-(1+i_0)^{-(t_6-t_4)}\right]}{i_0\cdot(1+i_0)^{t_4}}. \quad (9)$$

4. Experimental part

Equation (9) allows visualizing the effect of individual project parameters on the increase in net present value. The data for an illustrative example are shown in table 1.

Table 1. Data of IIP parameters

r, %	CK_1, million rubles	$3K_2$, million rubles	u_{t}, million rubles	t_1, years	t_2, years	t_4, years	t_5, years	t_6, years	t_7, years
10	50	100	5	3	5	7	8	9	10

Functional dependence $\Delta NPV_{\text{t}} = f(i_0)$ built according to the table 1 is presented in Figure. 2. It schematically displays the general tendency for the increase effect of the project's net present value to change: with an increase in the rate of return, the net present value increases, at a certain value it reaches a maximum and then begins to decrease, which corresponds to the well-known dependence of *NPV* the decrease with the growth of the rate of return. Moreover, in numerical value, conditions (5), (6) and (7) of clause 3 of this work are fulfilled.

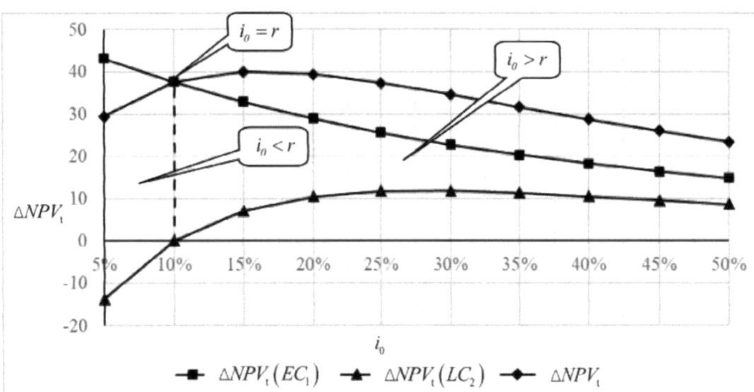

Figure 2. Impact of the rate of return i_0 on increase ΔNPV_{t} for IIP

The critical value of the rate of return i_0, in which the increase effect decreases to zero and then becomes negative, can obviously be found using a numerical solution of the nonlinear equation

$$\frac{EC_1}{(1+i_0)^{t_1}} - u_{\text{t}}\cdot\frac{1-(1+i_0)^{-(t_7-t_5)}}{i_0\cdot(1+i_0)^{t_5}} + \frac{LC_2}{(1+i_0)^{t_2}} - \frac{LC_2\cdot r\cdot(1+r)^{(t_4-t_2)}}{[1-(1+r)^{-(t_6-t_4)}]}\cdot\frac{\left[1-(1+i_0)^{-(t_6-t_4)}\right]}{i_0\cdot(1+i_0)^{t_4}} = 0$$

regarding i_0, for example, using the MS Excel module "Parameter Selection".

The influence of the loan rate r (with $i_0 = 15\%$ and the data of Table 1) on the increase of the NPV indicator is shown schematically in the graph $\Delta NPV_t = f(r)$ shown in Figure 3.

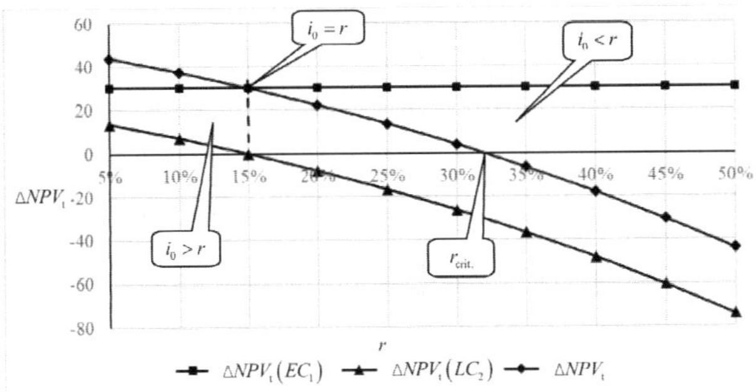

Figure 3. Impact of the loan rate r on increase ΔNPV_t for PV

With an growth in the loan capital (ceteris paribus), the positive effect of the increase begins to decline, as payments to repay the loan grow, and the contribution $\Delta NPV_t(EC_1)$ to the increase of the net present value of the project, as follows from (9), does not change. Moreover, in numerical value, conditions (5), (6) and (7) of clause 3 of the present work are also satisfied.

The critical value of the loan rate $r_{crit.}$, at which the increase effect decreases to zero and then becomes negative, can be found by numerically solving equation (9) relative to using the MS Excel module "Parameter Selection". For the IIP data given in Table 1 and $i_0 = 15\%$, the solution of such an equation gives a result $r_{crit.} = 33{,}24...\%$ that fully corresponds to the result obtained in Figure 3.

The influence of the time t_2 of introducing loan capital LC_2 into the financial-time diagram for the implementation of IIP (with $i_0 = 15\%$ the data of Table 1) on the increase of the NPV indicator is displayed on the graph $\Delta NPV_t = f(t_2)$ shown in Figure 4. The later the capital LC_2 is introduced into the IIP implementation scheme, the less, as follows from (9), its positive contribution $\Delta NPV_t(LC_2)$ to the project NPV increase. The contribution $\Delta NPV_t(EC_1)$ to the increase of the project's net present value does not depend on time t_2, therefore (ceteris paribus) the positive increase effect begins to decline.

The critical value of loan capital entry time $t_{2crit.}$, at which the increase effect decreases to zero and then becomes negative, can be found by numerically solving equation (9) with respect to using the MS Excel module "Parameter Selection".

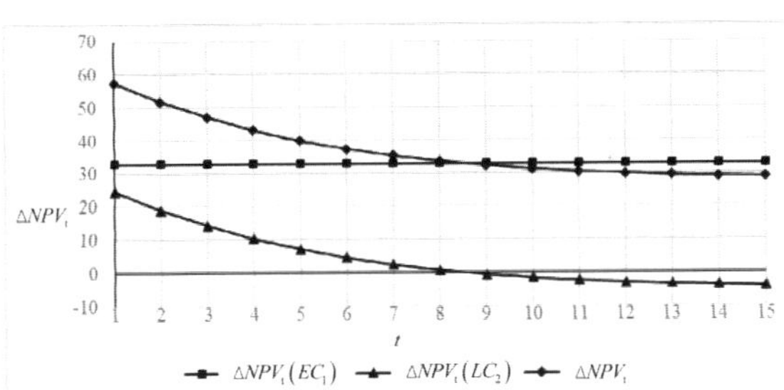

Figure 4. Impact of the entry time effect LC_2 on increase ΔNPV_t for IIP

5. Conclusion

The use of loan capital in the economy in general and in the implementation of innovative projects in particular is both an urgent task in terms of its theoretical significance and contradictory in terms of its practical implementation. On the one hand, attracting borrowed capital enables investors to implement promising innovative projects in the event of a shortage (or lack of own funds). On the other hand, the need to repay debts reduces the positive effect of obtaining a project's net present value and, thereby, reduces its financial stability. The latter circumstance is especially important to consider in the context of inflation and unstable market conditions.

The way out, in our opinion, is the construction of mathematical models of performance indicators in the form of functions of project parameters that will allow accurate determination of the conditions for obtaining the effect of the increase in net present value of the project when using both loan and equity in the form of mathematical relationships between the individual parameters of the project. In this work, this was done for the net present value indicator. The results obtained demonstrate the possibilities of mathematical modeling for studying the dynamics of investment processes with a view to their further application in practice.

References

[1] The official website of the Government of Russia 2011 *The strategy of innovative development of the Russian Federation for the period until 2020,* available at: http://government.ru/docs/9282

[2] Ed. Ilyenkova S 2012 *Innovation Management: A Textbook for High Schools* (Moscow: UNITY-DANA)

[3] Ed. Barysheva A 2015 *Innovation Management: A Study Guide* (Moscow: Publishing and Trading Corporation "Dashkov and Co.")

[4] National Innovation Report in Russia 2017, available at: https: // www.rvc.ru/upload/iblock/c64/RVK_innovation_2017.pdf

[5] The 2017 Global Innovation 1000 study, available at: https: // www.strategyand.pwc.com/gx/en/insights/innovation1000.html

[6] Kirillov Yu, Dragunova E and Kravchenko A 2017 *Economics and management control systems* **4.2 (26)** 245 - 255

[7] Chetyrkin E 2010 *Financial Mathematics: A Textbook for High Schools* (Moscow, Delo)

IOP Conf. Series: Materials Science and Engineering **843** (2020) 012019 doi:10.1088/1757-899X/843/1/012019

Regulation and stimulation of engineering innovation processes at machine-building enterprises

O V Milekhina and I B Adova

Novosibirsk State Technical University, 20, K. Marx Ave., Novosibirsk, 630073, Russia

E-mail: olga.milekhina@gmail.com

Abstract. The paper is devoted to the substantiation of certain aspects of the system of regulation of activity and remuneration of process engineers. The research is based on a theoretical model for justifying management decisions based on comparing the costs and benefits of innovation. The object of observation is a machine-building holding within the state Corporation ROSTEC (Russia). Data from self-observations of the working hours of process engineers, their processing and personalized analysis in two projections (ordinary and innovative work) led to the conclusion that there is a resource of time for solving innovative problems and creating breakthrough technologies. The results obtained are taken into account when designing the process of forming remuneration for innovations, which includes three phases. The authors conclude that a well-founded system of regulation and stimulation of innovative labor processes makes it possible to significantly advance in the direction of activation of innovation.

1. Introduction

The results of research by economists show that the most important intangible assets of organizations are people, who are the heart of any innovation ecosystem [1]. People maintain a high level of personal creativity [2], form project teams to solve non-trivial tasks using the tools of the Fourth industrial revolution [3], produce and apply new knowledge in practice [4], and support an innovative corporate culture.

The ability to make management decisions depends on the involvement of employees in technological, marketing, organizational and environmental innovations. Statistical estimates of the total level of innovation activity in industrial production in Russia fluctuate at the level of 10.5-11.1% during 2006-2017 [5]. This is a low level that shows the presence of systemic problems in the organization of innovation processes. European countries show a higher level of innovation activity: Germany-61.5%, Belgium-52.8%, Finland-50%, France and Austria-41-43%, Bulgaria-21.6%, Poland-17.7%, Hungary-17.5% [6]. Manufacturing industries in Russia on average show a higher innovation activity of 21.3%, Only one in five (19.9%) machine-building enterprises takes part in technological, marketing and organizational innovations [5]. The article discusses two barriers to innovation in production activities – overloading employees with routine and lack of incentives. Their minimization will solve some of the qwerty effects and stimulate innovation.

The purpose of the article is to present the results of a personalized analysis of certain aspects of the system of regulation and remuneration of process engineers of the machine -building holding, which is part of the state Corporation ROSTEC.

The object of observation was one of the machine-building enterprises in Novosibirsk (Russia). The company provides 75% of the market for its products in Russia. Further expansion of the market is limited to competitors. This was a prerequisite for the search for a new market niche in the format of product diversification, namely, an engineering center for the provision of high-tech services . The corporate competence of the new center is provided by the professional competence of the plant's engineers.

2. Background Results and Discussion

The participation of engineers in innovation activities is determined by their personal needs and a set of information about the expected costs and benefits of innovation [7, 8]. This decision to support innovation will look like this [9]:

$$D = \begin{cases} 1, if \sum_{n=1}^{N} r_{n\,sub} \leq \sum_{k=1}^{K} r_{k\,sub.\,\mathrm{ex}} \\ 0, if \sum_{n=1}^{N} r_{n\,sub} > \sum_{k=1}^{K} r_{k\,sub.\,\mathrm{ex}} \end{cases}$$

where D is the the decision to support / resist processes (can take the value 1 or 0);
$r_{n\ sub}$ are the subjectively estimated resources of the n-th type spent on implementing a solution to support/resist innovation;
$r_{ik\ sub.\ ex}$ are subjectively estimated expected benefits of the k-th type that can be obtained from the decision to support/resist innovation D.

If innovations affect the entire workforce, then active communication in the group leads to the dissemination of information about the expected changes and the formation of the collective's total expectations at time t of the total expectations of the team [9]:

$$I_{ex}(t) = \sum_{m}\sum_{j} i_{\mathrm{ex}\,mj}\,(t)^{\circ}\gamma_{mj}$$

$$\gamma_{mj} = \begin{cases} 1, & if\ \tau_{mj} \leq t - t_{incl.\,j} \\ 0, & if\ \tau_{mj} > t - t_{incl.\,j} \end{cases}$$

where $I_{ex}(t)$ are the total expected benefits of innovation by the team of engineers at the moment t;
$i_{ex\ mj}(t)$ are the expected benefits of the m-type by the j-m engineer;
γ_{mj} is the inclusion multiplier for the benefits of the m-type j-th engineer (can take the value 1-if the engineer considers the benefit as a benefit from participating in innovations, 0-otherwise;
τ_{mj} is the waiting period for the benefits of m-type j-th engineer;
$t_{incl\ j}$ is time when an engineer is included in the innovation process (the beginning of innovation development).

Systematic and consistent work with engineers begins with the regulation of activities for solving and supporting ordinary tasks and technologies. Creating a model for performing work at different levels of detail will allow organizing techniques and methods for solving individual tasks in the workplace of engineers, establishing quantitative characteristics of time spent on working with ordinary tasks and processes and evaluating the quantitative characteristics of time spent on working with innovative tasks and processes.

The accumulated experience of regulating the operations of ordinary technological processes does not allow determining the norms of time for innovative tasks . For specialists of intellectual labor, for

IOP Conf. Series: Materials Science and Engineering **843** (2020) 012019 doi:10.1088/1757-899X/843/1/012019

example, engineers of the services of the chief engineer of industrial organizations, «the system of microelement labor rationing is not applicable. The specifics of this category of employees should be taken into account. New approaches must be found» [10].

The task of determining the time spent on the work of engineers with ordinary and innovative tasks and processes was updated during the implementation of the project to change the remuneration system for the personnel of the selected observation facility.

The analysis of business processes made it possible to conclude that the engineers of the chief engineer's service perform the following types of work efficiently, namely, providing a full cycle of research and development work, development of design documentation, development of new equipment, technologies, parts, products, improvement of existing production technologies and preparation of comprehensive technological documentation, ensuring the safety of production and the environment.

3. Results and Discussion

Using the method of working day photography, the actual time spent on the work of engineers in the chief technologist's Department with ordinary and innovative tasks and processes was measured. For two months (nine work weeks) the process engineers kept diaries of self-observation. Process engineers measured and recorded the time spent on innovative and ordinary tasks and processes. In this way, we obtained 45 data sets for seven jobs of process engineers, which recorded the actual time spent on working with ordinary and innovative tasks and processes.

Calculation of arithmetic mean values , median and mode of available samples (Table 1) allowed concluding that the values obtained for each series do not differ significantly, and, therefore, the distributions can be considered normal.

Table 1. Results of statistical calculations

Working places of the chief technologist's service	The ordinary tasks and processes		The innovative tasks and processes	
	Mathematical expectation, min	Standard deviation, min	Mathematical expectation, min	Standard deviation, min
The head of the group of rationing of materials	419,47	39,03	58,44	15,18
The engineer-technologist of the of the group of rationing of materials and new equipment	454,40	9,67	25,60	9,67
The leading engineer-technologist of the of the group of rationing of materials and new equipment	456,80	7,76	23,20	7,76
Engineer-technologist for new technology	420,80	19,33	59,07	12,98
The engineer-technologist of the new equipment	444,89	13,52	36,27	9,96
The leading engineer-technologist	420,80	12,75	59,96	12,29
The leading process engineer in the technology Bureau	336,62	21,24	143,29	18,50

The deviation of the total duration of observations by day is less than 1%.

IOP Conf. Series: Materials Science and Engineering **843** (2020) 012019 doi:10.1088/1757-899X/843/1/012019

Processing of the observation results allowed concluding that the working time of engineers is mostly spent on working with ordinary tasks and processes (figure 1).

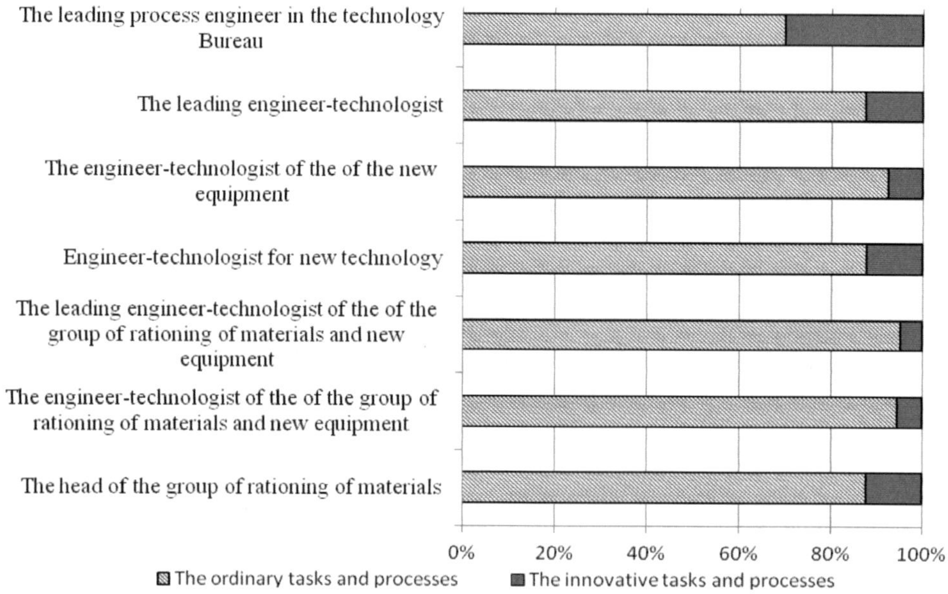

Figure 1. Allocation of time spent by process engineers for ordinary and innovative tasks and processes.

This ratio of time spent does not allow technology engineers to focus on achieving the goals of the company's technology policy, stimulate the development of personal professional competencies, and create and implement innovations.

Regulation of ordinary tasks and technologies provides an opportunity to delegate authority and transfer a significant part of it directly to production units. In this case, the chief engineer's service engineers have a resource of time to solve innovative problems and create breakthrough technologies, can concentrate on developing plans for medium-term innovative development, and effectively manage innovations in divisions.

At the second stage of the research, the analysis of innovation incentives was carried out and the design of the process of forming the incentive part of remuneration for innovation was carried out. Initially, the amount of remuneration for process engineers consisted of the base part (salary), compensation payments (for example, for the danger and harmfulness of work) and the bonus part of remuneration. Initially, the award was not of a stimulating nature - there were no relevant parameters for evaluating the intellectual work of technologists. The system of the key performance indicators (KPI) was proposed in the development of the existing incentive system for the enterprise with a focus on results.

It was suggested to use two projection of labor productivity assessment such as operational (for evaluating ordinary processes) and strategic (for evaluating innovative processes). The strategic component of the KPI should ensure the involvement of enterprise specialists in the process of achieving the main goals of any business is the attracting and retaining consumers (customers), developing employee competencies across the entire palette of competencies (hard skills, soft skills and digital skills), diversifying sources and increasing revenue, reducing costs.

The authors proposed to form a stimulating part of the remuneration in three stages sequentially (Figure 2): designing of the KPI system, planning of the indicators for the upcoming period, assessment of the performance by KPI for the reporting period and calculation of bonus remuneration.

Input and output information flows ensure the execution of the proposed process.

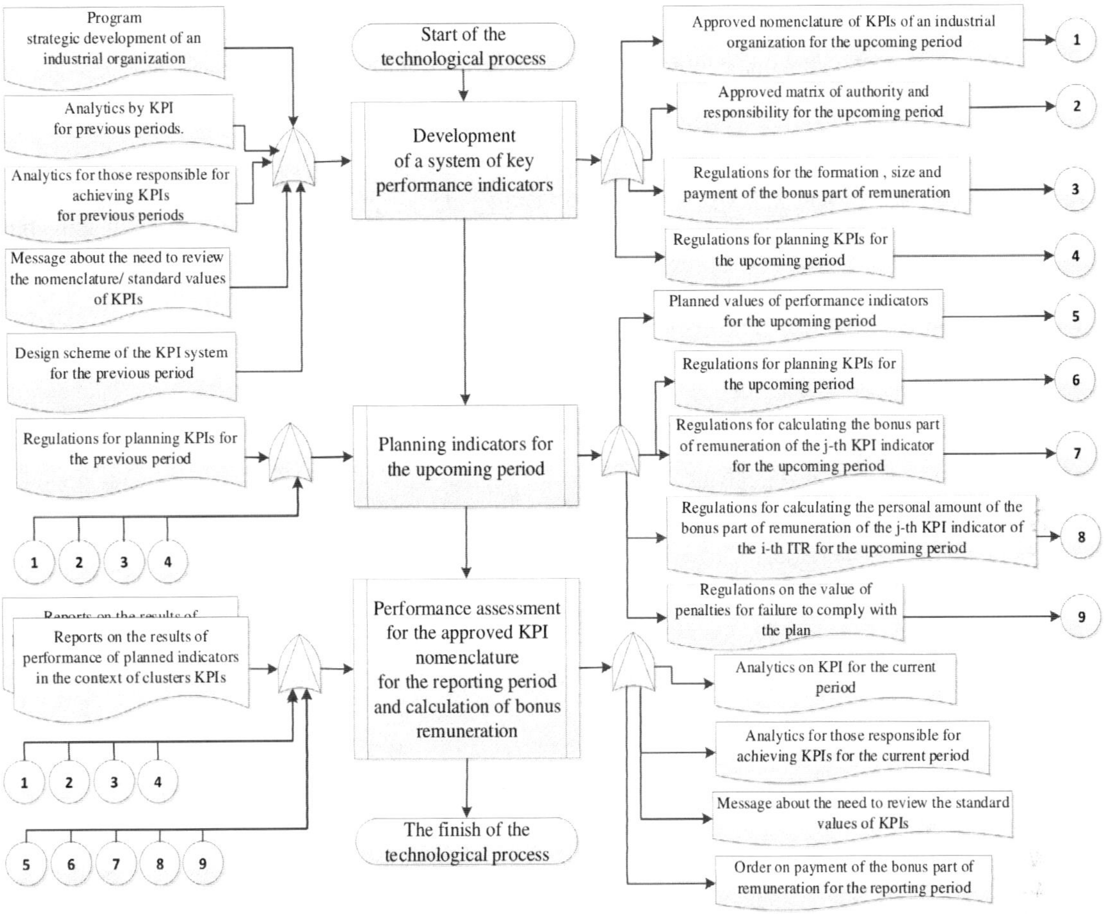

Figure 2. Process of designing the incentive part of remuneration for employees of the enterprise

4. Conclusion

The results of the study clearly show that the regulation and stimulation of innovative labor processes of enterprise engineers can have an additional influence on the decision to include engineers in innovation activities. This will significantly improve the company's innovation climate. Today, the current activity of engineers involves solving too many ordinary tasks. Engineers have limited capacity to maximize the productivity of their intellectual assets. It is necessary to stimulate engineering innovation processes in machine-building enterprises and redistribute the time spent in favor of working with innovative tasks and processes. This will lead to increased production of technological, marketing, organizational and environmental innovations and an overall increase in the efficiency of all production processes of the machine-building enterprise.

References

[1] Agisni S, Kurniawati R, Ambarsari A, and Andrawina N. 2019 Competency Measurement Instrument Design for Maintenance Staff of Electronic Expertise with SECI MethodIOP *Conference Series: Materials Science and Engineering* **528 (1)**

[2] Endovitskaya E V 2020 Assessment of the Company's Staff Creativity as the Basis for Their Adjustment to the Terms of the Digital Economy *Lecture Notes in Networks and Systems* 489-498

[3] Huang Z, Kim J, Sadri A, Dowey S, Dargusch M S 2019 Industry 4.0: Development of a multi-agent system for dynamic value stream mapping in SMEs *Journal of Manufacturing Systems* **52** 1-12
[4] Kong F 2019 Development of metric method and framework model of integrated complexity evaluations of production process for ergonomics workstations International *Journal of Production Research* **57(8)** 2429-2445
[5] *HSE University. Research Departments. Institute for Statistical Studies and Economics of Knowledge.* Official website, available at: https://issek.hse.ru/press/194804026.html
[6] *Rosstat.* Official website, available at: https://www.gks.ru/folder/14477
[7] Spieth P, Roeth T and Meissner S 2019 Reinventing a business model in industrial networks: Implications for customers' brand perceptions *Industrial marketing management* **83** 275-287
[8] Loch C and Sommer S 2019 The Tension Between Flexible Goals and Managerial Control in Exploratory Projects *Project management journal* **50 (5)** 524-537
[9] Samosudov M V 2012 Theory of corporate interaction as an analysis tool in the process corporate governance *Management science* **2** 24-30
[10] Podverbnyh O E, Mezhova I A and Sokolova E L 2019 Approaches to the unification of labor standards high-tech specialists professions *Russian Journal of Labor Economics* **4** 1343-1352

IOP Conf. Series: Materials Science and Engineering **843** (2020) 012020 doi:10.1088/1757-899X/843/1/012020

Optimization of costs of residual deficit regulation in production system

V I Mamonov and O V Milekhina

Novosibirsk State Technical University, 20, K. Marx Ave., Novosibirsk, 630073, Russia

E-mail: olga.milekhina@gmail.com

Abstract. The article considers the production subsystem as an element of the technological chain of a production system of any structure. Improvement of business efficiency involves optimizing key integration processes and coordinating the interaction of subsystems and links in technological chains. Each subsystem participates in the production of final products and is both a consumer and a supplier of products. The influence of various factors on the production process, including random ones, is the cause of violations of the time standards for product movement, which is expressed in deviations of the actual timing of the launch and release of planned accounting units from the planned standards. The concepts are introduced: late start-up in the subsystem - input deficit, delay transfer to the time of planned start - up in a neighboring subsystem along the technological chain-residual deficit. The solution of the subsystem is analyzed taking into account the amount of deficit transmitted along the technological chain. The function of costs for regulating external instability is defined, which includes two parameters: the value of the input deficit and the value of the residual deficit. If there is a residual deficit, compensation for the cost of reserving a neighboring subsystem along the chain is carried out in accordance with the linear function of the penalty and should be sufficient to neutralize it in this subsystem. A solution to the problem was obtained and a policy for regulating the remaining deficit was formulated.

1. Introduction

The studied production systems have a complex multi-level structure with a huge number of economic, organizational and communication links [1], so the stability of their functioning depends on the activity of their constituent elements, each of which in turn is subject to various disturbances that are random in relation to a separate division. The connection of elements of a production system can be characterized by a different structure: it can be a linear straight-through technological chain, with elements of a fan structure both at the input of a separate link in the system and at the output. The main structural element of the production system is always the production link (subject-closed or technological section, group or separate work centre, group of jobs). In the article, the object of consideration is a link (subsystem) of the production system.

In accordance with the structure of production links, each subsystem, having completed the finished part of the production process, transfers products to the next "supplier – consumer" subsystem in the specified time frame. Under the conditions of random disturbances (both external and internal) in discrete production, which are random in relation to a separate subsystem, there is a mismatch between the actual delivery dates of products and the planned ones. Because in the actual production

IOP Conf. Series: Materials Science and Engineering **843** (2020) 012020 doi:10.1088/1757-899X/843/1/012020

to coordinate the supply of products is possible only up to a certain time interval, then the supply of "just in time" is understood as adopted in the current system of operational production planning time period - the time interval coordination of product delivery just in time. This standard shows the accuracy (just in time) with which production units set delivery times for products to adjacent production subsystems in the process chain. Thus, if production units ensure the receipt of products in the supplier-consumer chain within the approval interval, then we can assume that the system meets the specified criteria for efficiency and reliability [2, 3, 4, 5, 6].

2. Materials and methods

The main regulatory document for drawing up calendar schedules for the movement of planned accounting units (PUS) for process operations is the launch-release schedule. In figure 1 a fragment of the launch-release schedule of two adjacent batches for a specific name is given. The schedule shown in the figure for the launch and release of batches characterizes the operation of production system (a subject-closed section, work centers, or a group of jobs) with a different or uniform rhythm for all parts batch names - R_0. The planned and actual dates of launch and release of partsD_{jp}^{pl} , D_{jp}^{al}, D_{jp}^{pr}, D_{jp}^{ar} of the j-th name are indicated by the index p – the serial number of the batch; the difference between the dates(D_{jp}^{pl} - D_{jp}^{al})is equal to the planned duration of the production cycle of manufacturing parts in the production system.

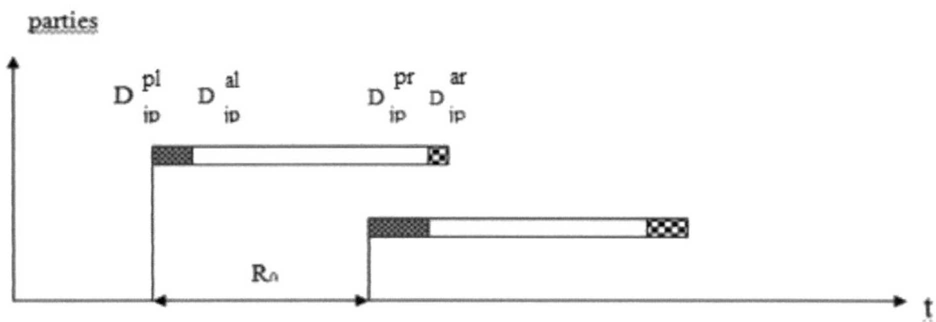

Figure 1. Planned and actual schedule start of production of batches of parts

Due to random factors, both internal and external, as well as objective reasons due to production conditions, the actual execution of the launch-release schedule for this fragment is determined by the actual launch and release dates of parts batches. The shaded fragments represent, respectively, the delay in delivery of the planned accounting unit from the production subsystem adjacent to the process chain, which is equal to the difference between the actual launch date of the PUS and the planned date (D_{jd}^{pl} - D_{jd}^{al}) and which is the input deficit; the difference between the actual release date of thePUS and the planned date (D_{jd}^{pr} - D_{jd}^{ar}) is the residual deficit.

In theconditionsofstructuralandtechnologicalconnectivity of subsystems by the sequence ofproduction processes performed, the delivery of products in the "just in time" mode is regulated by direct methods of production process control. However, in order to be more effective, direct methods of production management are always supplemented by methods of economic regulation, the focus and content of which is ultimately reduced to taking into account the economic interests of subsystems. The organizational form of such accounting is the system of material remuneration and compensation for damages[2, 7].This system provides a function of stabilizing planned standards for the implementation of the production process by embedding the economic interests of subsystems in direct management methods. With an effective system of economic regulation, the effect of self-regulation is observed, i.e. ensuring a stable flow of the production process by neutralizing deviations from the supply of products between subsystems. This management principle is important in terms of pulling production and ensuring its competitiveness through rapid response [8].

Based on the statistical model of the movement of labour items over time along the "supplier – consumer" technological chain, the instability of the subsystem can be described using average delays in the delivery of products to related units [2, 9]. It is the size of these delays that characterizes the level of redundancy, as well as the size of the losses that occur in the division that consumes products. Losses from instability determine the amount of additional costs in other divisions throughout the supplier-consumer chain. These costs perform a compensatory function for their intended purpose and are intended to completely or partially neutralize the consequences of instability, which is transmitted through technological links to production system units. The amount of costs for the full or partial elimination of the consequences of instability is determined by the level of reserves created in each production unit, as well as between them. The most used methods of reservation in practice are volume and temporary reservation [10, 11, 12].

Further consideration of the behaviour of the production link of the system requires determining the values of input and residual deficits. Deviations of the actual moments of launching batches into production from the planned ones are random values. Let us denote the average duration of the delay (input deficit) in the production unit for the set of planned accounting units through θ_0.

Note that there may be several approaches to determining the average value of the input deficit. Let all the set of names of planned accounting units to be processed in the production link in the considered period be equal to m. The actual start date of production of a planned accounting unit in the production system can be deferred from the planned date for a time multiple of the planned control period - θ_{0i}.

We introduce a random variable $\theta^0 = \max(\theta_{01}, \theta_{02}, ..., \theta_{0m})$, where θ_{0i} is a discrete random variable that takes the values $\theta_j^0 = j$, $j = 1(1)N$ with a step equal to the value of 1kpp and with statistical probabilities p_{ij}, $i = 0,..., m$.

The length of the time interval for approving the delivery of a batch on time may vary depending on the production conditions and the operational scheduling system. For example, 1kpp = 0.25 the duration of the work shift or 1 kpp= 0.5 the duration of the work shift, etc. We calculate the probability values and get the distribution series of a discrete random variable θ^0 [13] :

$$P\left(\theta_0 = \theta_j^0\right) = \sum_{i=1}^{m-1} P\left(i, \theta_j^0\right) + P\left(\theta_{0i} = \theta_j^0,\ i = 1,..., m\right),\ j = 1,..., N,$$

where

$$P\left(i, \theta_j^0\right) = \sum_{v_1=1}^{m-i+1} \sum_{v_2=v_1+1}^{m-i+2} \cdots \sum_{v_i=v_{i-1}+1}^{m}$$

$$P\left(\theta_{0v_1} = \theta_j^0, \theta_{0v_2} = \theta_j^0, ..., \theta_{0v_i} = \theta_j^0, \theta_{0\mu} < \theta_j^0,\ \mu \in \{1,..., m\} \setminus \{v_1,..., v_i\}\right).$$

The description of random variables as discrete is explained by the greater adequacy of the process under consideration. Using continuous random variables in calculations is more convenient, but it requires justifying the accepted types of distributions and obtaining their convolutions. We also believe that the assumption of independence of random variables is justified by the required accuracy of calculations for practical purposes.

The average delay duration (input deficit) in the production unit for the set of planned accounting units is defined as the mathematical expectation of a random variable θ^0:

$$\theta_0 = \sum_{j=1}^{N} P(\theta^0 = \theta_j^0) \cdot \theta_j^0 \cdot$$

Similarly, the average value of the residual (output) deficit is determined. The availability of statistical data is ensured by the simulation of calendar schedules, taking into account the provision of additional time to the production link (temporary redundancy) to reduce its own instability and compress the input deficit.

We introduce into consideration the cost function for the regulation of external instability. The division, being an intermediate link in the technological sequence of manufacturing the final product, is a source of delays in deliveries to the planned date, that is, it operates with an average output deficit. The process of restoring consistency, which consists in partially or completely neutralizing the input deficit, requires additional labour and material costs in the division. By its content, the costs incurred are the costs of regulating external instability - Z. The amount of these costs is a loss to the consumer, and therefore must be reimbursed in the form of penalties imposed on it, either by the unit that is the source of the deficit, or by the system.The amount of the penalty is the amount of expenses that the consumer subsystem needs for its own reservation in order to neutralize the residual deficit. We assume that for the purposes of the study, it is sufficient to determine the costs by a function that depends on two parameters – the value of the input and output deficit. For each subsystem, the cost function for regulating external instability has the form $Z(\theta_0, \theta)$, where the residual deficit of the previous subsystem is an input deficit to the next subsystem[14].

The residual deficit is the sum of two components, the first of which is the intrinsic instability of the subsystem, and the second is the input deficit allowed by the preceding subsystems. The simulation results show that the residual deficit is a positive value. With a relative balance of equipment group throughput and low load factors, the minimum residual deficit is small and may not be taken into account when reconciling the supplier-consumer production relationship over time. However, as the load factors increase and the input deficit remains at the same level, the minimum possible amount of residual deficit increases because it is determined by the smaller amount of reserves remaining at the disposal of the division.

3. Results and Discussion

The cost function for regulating external instability is a function of two variables $Z(\theta_0, \theta)$. The function reflects the variety of applications and the extent to which internal regulators are used to compensate for instability in the division. The general nature of the dependence of costs on variables shows that the cost function for regulating external instability has a clearly expressed nonlinear form.

When analyzing a function in order to further approximate it with simpler functions, it is most important to establish a relationship between the growths of the function of regulation costs when the residual deficit decreases. We will approximate the function of regulation costs assuming that the relative increase in costs per unit of output deficit reduction is constant

$$\left[dZ(\theta_0, \theta) / Z(\theta_0, \theta)\right] / \left(d\theta / \theta\right) = -\alpha . \tag{1}$$

Solving equation (1), using natural boundary conditions and taking into account that the minimum cost of completely neutralizing the consequences of the deficit itself depends on the value of the input deficit, allows obtaining a function of the control costs depending on the value of the input and residual deficit in the form of an expression

$$Z(\theta_0, \theta) = \vartheta \theta_0^{1+\alpha} \left(c + \theta\right)^{-\alpha}, \quad \alpha > 0. \tag{2}$$

It is obvious that the amount of expenses allocated in the division to regulate the residual deficit, which is determined by the formula (2), takes into account the most significant requirements. If the values of the residual deficit at the output of the production system are equal, then the value of the costs of regulation will be greater if the input deficit is greater.Meaning, if the residual deficit is equal to zero, it means that the input deficit is fully compensated and the control cost function must take the maximum value

$$Z(\theta_0,0) = \max Z(\theta_0,\theta).$$

An exogenously set parameter ϑ,equal to the cost of fully compensating the input deficit unit, a constant value of c, and an exponent α allow flexible change the shape of the lines of the cost function and thus ensure that they correspond to the actual values. Parameter c corresponds to the amount of the costs of regulation that is sufficient to completely neutralize the input deficit in size θ_0.

The linear penalty model is used in the subsystem interaction model in order to establish relationships between the sizes of mutual liability claims. For a linear function of theresidual deficitpenalty $w(\theta) = \beta \cdot \theta$, the required minimum of the amount of the costs of regulationand penalties is determined by the expression:

$$W(\theta_0,\theta) = \min_{\theta}\{Z(\theta_0,\theta) + \beta\,\theta\} \text{ if } \quad \theta_0, \theta \geq 0. \tag{3}$$

Finding the minimum of the function gives the following result [15]. When searching for valid values, the equation is

$$\frac{dW(\theta_0,\theta)}{d\theta} = 0, \text{ if } \quad \theta = 0.$$

So

$$-\alpha\vartheta\theta_0^{1+\alpha}(c+\theta)^{-\alpha-1} + \beta = 0, \quad \theta = 0,$$

wherefrom

$$c + \theta = \left(\frac{\alpha\vartheta}{\beta}\right)^{\frac{1}{1+\alpha}}\theta_0 = k\theta_0 \quad (4)$$

and with a sufficient deficit equal to zero, we get the value of the input deficit $\tilde{\theta}_0$ such that for all the value $\theta_0 \leq \tilde{\theta}_0$ of the function (3) is the minimum for

$$\tilde{\theta}_0 = c/k. \quad (5)$$

For input deficit values $\theta_0 > \tilde{\theta}_0$, the optimal value of the deficit $\hat{\theta}$ passed to the consumer unit is found from the solution of the equation

$$\frac{dW(\theta_0,\theta)}{d\theta} = 0, \quad \text{if } \theta \geq 0, \theta_0 \geq \tilde{\theta}_0.$$

where the output deficit value is calculated as

$$\hat{\theta} = k\theta_0 - c = k(\theta_0 - \tilde{\theta}_0).$$

By combining the results we have obtained

$$W(\theta_0,\theta) = \begin{cases} (\beta/\alpha)\theta_0^{1+\alpha}\tilde{\theta}_0^{-\alpha}k = W(\theta_0,0), & \theta_0 \leq \tilde{\theta}_0 \\ (\beta/\alpha)\theta_0 k + \beta \cdot k(\theta_0 - \tilde{\theta}_0) = W(\theta_0,\hat{\theta}), & \theta_0 \geq \tilde{\theta}_0 \end{cases} \tag{6}$$

The general view of the function, which is monotonically increasing on each of the segments, is illustrated in Figure 2.

The optimal regulatory policy of the value of the remaining deficit is formulated on the basis of the result (Eq. 6): if the value of the input deficit in the subsystem is within the range of $0 < \theta_0 < \tilde{\theta}_0$, so it is economically advantageous for the subsystem to eliminate the deficit completely using its own resources and reserves. With an input deficit $\theta_0 > \tilde{\theta}_0$, the optimal policy is to partially compensate for the deficit.

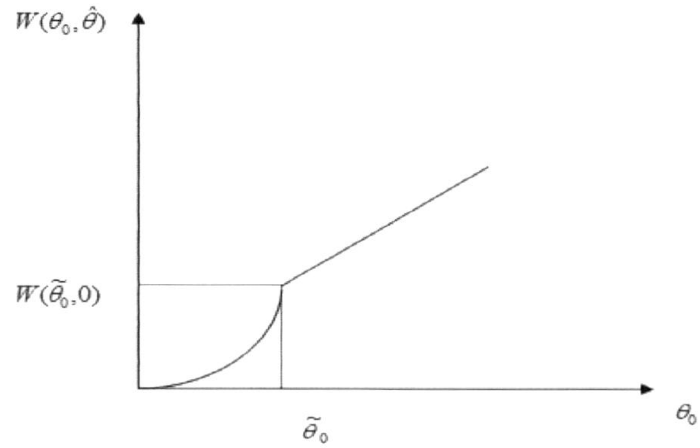

Figure 2. Graph of the cost function for regulating the external instability of the production system

The compression ratio of the deficit is a characteristic of the level of redundancy of the production system:

$$\mathrm{CRD} = \begin{cases} 1, & \theta_0 = \tilde{\theta}_0, \\ 1 - k\left[1 - \tilde{\theta}_0 / \theta_0\right] = 1 - \left[k - c/\theta_0\right], & \theta_0 > \tilde{\theta}_0 \end{cases} \tag{7}$$

Depending on the ratio of constants that determine the coefficient k, the residual deficit may increase at a higher or lower rate relative to the increasing deficit at the input. For known values k and c for the value of the input deficit from expression (7), the deficit compression coefficient corresponding to the minimum amount of regulation costs and penalties is easily determined.

4. Conclusion

The results obtained can be used when considering a sequential scheme of production relations "supplier-consumer" or sequential fragments in General schemes. In the case of a fan structure of links subsystems, the average input or residual deficit values are determined using a series of distributions across a set of supplies from previous subsystems or consumer subsystems. If the components of the cost management function are known, they are distributed among the subsystems in accordance with the accepted rules. The costs of controlling the residual deficit in an explicit form (6) allow them being used in tasks of optimizing the total costs of reserving production systems, as well as in managing supply chains [16].

Note an important circumstance. The concept of "reliability of the production system" and the related concepts of "redundancy", "stability", and "sustainability" make sense for stationary, relatively stable operating conditions of the production system [1, 2, 12, 17].If parametric product changes are significant enough that the system does not detect stationary conditions, then special adaptation reserves are needed for stable operation. The results of simulation experiments show that the residual deficit when the average values of the load factors in the system exceed 0.85 in many cases exceeds the value of the input deficit due to the inherent instability of the subsystem. The most effective tools for reducing the residual deficit in practice are: earlier product launches times compared to the planned dates in the production system (advance) and increasing the working time utilization rate. A significant reduction in the residual deficit and the cost of managing external instability is achieved by reducing the size of batches.

To determine the statistical characteristics of input and residual deficits in high-tech production, it is advisable to use the applied results of the theory of Queuing systems. The use of Queuing systems with management allows establishing an analytical relationship between the amount of residual deficits and the cost of reserving the subsystem [18].

References

[1] Grundig, C-G 2009 *Fabrikplanung: Planungssystematik, Methoden, Anwendungen* (München; Wien: Hanser)

[2] Lviv Yu A and Satanovsky R L 1984 *Intensification of machine-building production : organization and planning*(Leningrad: Mechanical Engineering)

[3] Sandler D 1966 *Technique of reliability of systems* (Moscow: Nauka)

[4] Gento A M and Redondo A 2001 *Reliability of production systems.* In:Handbook of Management under Uncertainty. Applied Optimization (Springer, Boston, MA)

[5] EgorovVN and KorovinDI 2006 *Fundamentals of the economic theory of reliability of production systems* (Moscow: Nauka)

[6] ChlebusM and Werbińska-Wojciechowska Sylwia2017 Issues on production process reliability assessment – review *Research in Logistics and Production* **6 (6)** 481–497

[7] Ashimov A A, Burkov V N, Djaparov B A andKondratev V V 1986*Coordinated management of active production systems*(Moscow: Nauka)

[8] Suri R 2010 *It's about time: the competitive advantage of quick response manufacturing*(Portland: Productive Press)

[9] Pervozvansky A A 1975 *Mathematical models in production management*(Moscow: Nauka)

[10] Cherkesov G N 1974*Reliability of technical systems with temporal redundancy* (Moscow: Soviet radio)

[11] Ryabinin I A and Cherkesov G N 1981 *Logical- probabilistic methods for studying the reliability of structurally complex systems* (Moscow: Radio and communications)

[12] Chuprov S V2007 *Management theory and stability of production systems* (Irkutsk: Publishing house BGUEP)

[13] Wentzel E S 1998 *The probability Theory* (Moscow: Higher school)

[14] Khayrullina M V and Mamonov V I 2017 *MT S* **243**

[15] Hadley G 1964 *Nonlinear and dynamic programming* (Addison-Wesley Publishing Company)

[16] Egorov V N, Chernova M V 2015 *News of higher educational institutions. Textile industry technology* **2 (356)** 5-10

[17] Mauergauz Yu 2016 *Advanced Planning and Scheduling in Manufacturing and Supply Chains* (Springer-Verlag)

[18] Mamonov V I and Milekhina O V 2018 Actual problems of electronic instrument engineering *(APEIE–2018)* **1** 106-110

IOP Conf. Series: Materials Science and Engineering **843** (2020) 012021 doi:10.1088/1757-899X/843/1/012021

Study conclusion mechanisms in hybridization of neural and logic intelligent systems

A M Pishchukhin and G F Akhmedyanova

Orenburg state University, 13, Pobedy ave., Orenburg, Russia, 460018

E-mail: pishchukhin55@mail.ru

Abstract. The paper investigates the integration problems neural networks sections allocated in accordance with the functions performed. As an initial example, we study two trained neural matrices, one of which performs the function of recognizing the position of the manipulator and its speed, and the second generates responses using the corresponding robot drives. In addition to training mechanisms based on the adjustment of transfer weights, the mechanisms based on the supply of excitation / inhibition waves from the edges of the neural network are also affected. At the same time, training is reduced to multiple forced guiding of the robot manipulator along the target path, during which the desired paths of beatings are automatically found using comparators. The matrices are trained separately and then joined using a standard signal generated by a special unit. The division of the neural network into two sections allows building the necessary logic between them. This hybridization significantly expands the range of tasks. As a test problem, the task of searching for a given part from a jointly located set of parts in general position was selected. At the same time, the control system finds a given part, rotates its predetermined model until a matching image is obtained, and thereby understanding how the part is located and how best to capture it occurs.

1. Introduction

At present, neural networks are often used in practice as an effective artificial intelligence tool [1]. This is caused the speed of processing a large amount of information due to parallel processes in them, as well as a wide range of tasks, the solution of which they can be trained with the required accuracy. However, in practice, networks that can be called single-stage are more often used, that is, the network is a kind of information processing channel, the initial information is input and the result appears on the output. Some studies nevertheless use sections of the network that are allocated according to a functional basis [2,3].

2. Theory

Hybridization of intelligent technologies is quite developed. However, it can be interpreted in different ways - for example, in [4] it is associated with a decision-making procedure by a diverse group of specialists. It can be related to the complexity of the tasks and can be very different when predicting the behaviour of time series based on a modular neural network [5] or detection of objects for image analysis based on convolutional neural networks proposed in 1989 by Yann LeCun, which gave a second life to neural network theory [6].

F. Rosenblatt, who proposed the idea of a perceptron, understood it as a universal model of the whole brain [7]. However, medical studies show that the brain is nevertheless divided functionally.

The possibilities of a functional approach in technology were studied in [3], where local neural networks were applied to simulate the process of annealing a zinc strip, to search for technologically rational temperature settings in the zones of the heating section and to form actions applied to the input of the actuator.

Another way to isolate local sections of the neural network is using the so-called convolutional networks. This use of a local receptive field for each neuron of the convolutional layer, the formation of convolutional layers in the form of a set of cards, the neural elements of which have the same synaptic connections, and the presence of subsampling layer cards, increase the network invariance to distortions [8-10]. The general architecture of the convolution network is shown in Figure 1.

The advantage of the "island" application of neural networks is the ability to insert logical elements between the islands, that is, it creates hybrid neural-logical networks, which greatly expands the range of tasks.

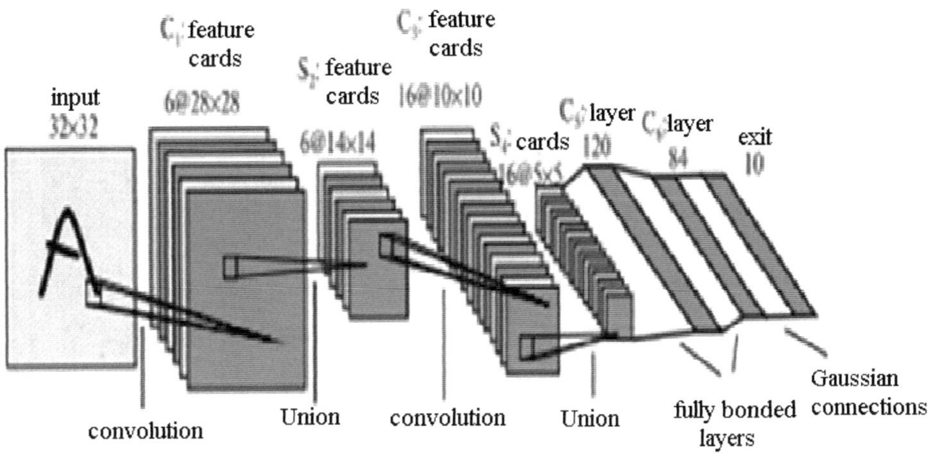

Figure 1. LeNet-5 Network Architecture [1].

3. Model

We study the functional separation of sections of the neural network into the matrices of trained neurons, used in [2] (Figure 2). Two matrices are distinguished in the device, one of which performs the functions of technical vision, it can be called recognizing (in Figure 2 on the left), the second, let's call it responsive (in t Figure 2 on right), implements responses, setting the magnitude and direction of movement of the manipulator using the drives. The device is first taught a matrix that implements the necessary reactions, forming the driving signals using a special block of standard signals (in the figure 2, number 18). Then, repeatedly repeating the necessary trajectory of the manipulator, the first matrix is trained, achieving at its outputs signals that are identical with the standard signals using comparators (in the figure 2, number 17), forming positive or negative reinforcements in training. Now the outputs of the recognition matrix can be connected to the inputs of the reacting one and a loop will be obtained in which both the given trajectory of the manipulator using neurons 8 and its speed along this trajectory are controlled due to the presence of differentiating chains in the synapses of the trained neurons 10. If there are any irregularities in the field of view of the recognition matrix, for example, foreign objects or an increase in environmental resistance (for example, clinging to any fabric, wire tension), the manipulator will be stopped or it will restart.

Another mode of operation of the device can be implemented if we create a set of all possible reactions of the manipulator in this scenario of behavior. All that remains is to train the recognition matrix so that it outputs exactly those signals that correspond to the manipulator reaction adequate in the given conditions. Such a system will realize conditional reflexes, as it were, in contrast to the first regime, where reflexes were formed unconditionally. It can be used to respond to detected defects in an appropriate control system [11].

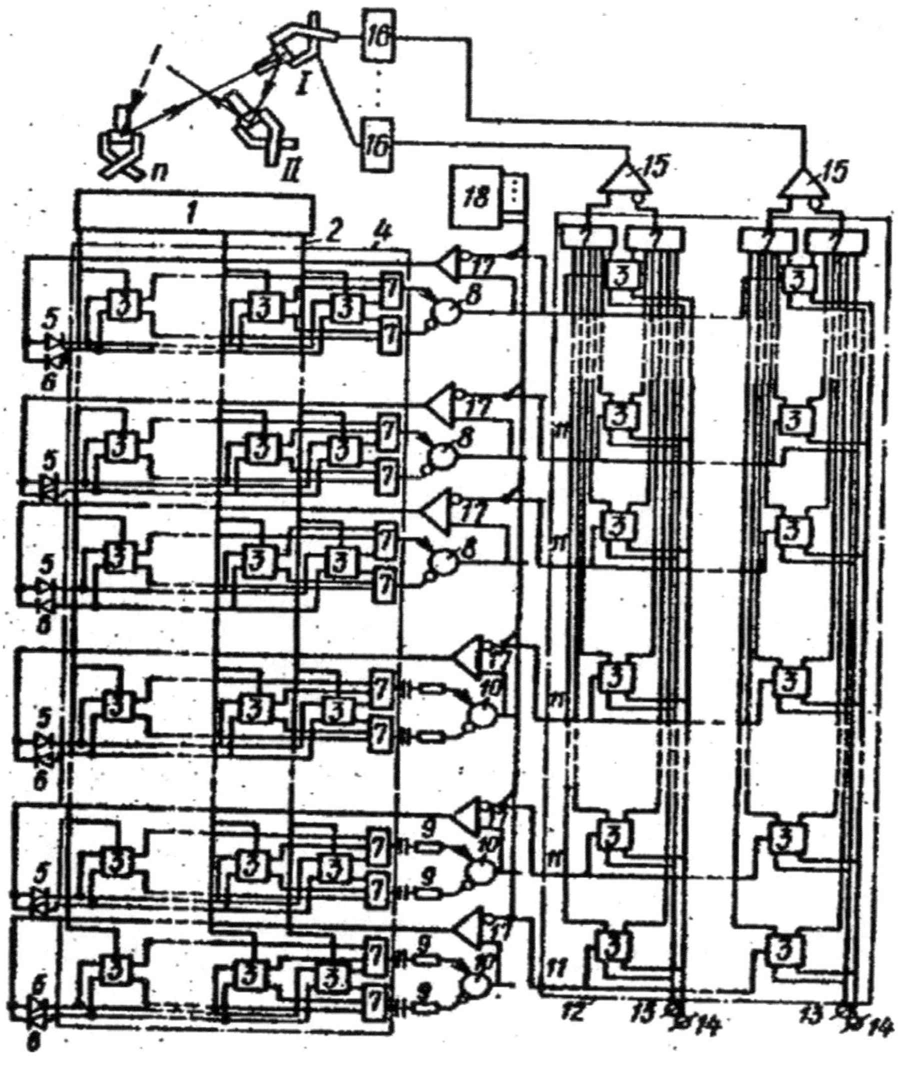

Figure 2. Device for nervous system modeling

4. Data and Method

We study the features of the application of the theory to the problem of choosing and capturing details of an arbitrary arrangement. Figure 3 shows a photograph of the pin (the extreme detail on the left), while the task is complicated by the presence in the field of view of other parts of similar shape.

The control system of the manipulator must recognize the necessary part, understand its position in space and choose the method of its capture.

Two types of processing are performed in parallel in the recognition matrix. In the first case, the idea of convolutional networks that increase the invariance of the device to distortion is used. As a result, the image of the pin is recognized and the position of its geometric center is determined. At the same time, noise is eliminated by filtering the image, including from glare, and the contours of the parts are determined. Then, the contours are completed at the points of their discontinuities, as is observed at the contour end of the pin shown in Figure 4.

To complete the contours, we can apply the mechanism of cellular logic. Moreover, each pixel of the contour image of the pin in Figure 4 corresponds to its own cell with simple response rules depending on the state of the neighbors. If the neighbors are in a state of inhibition, then this cell also goes into this state and vice versa in the case of excited neighbors. Cells corresponding to the existing contour lines stop the propagation of both excitation and inhibition waves.

Figure 3. Figure with short caption (caption centred).

Since there are often discontinuities in the contour when the signals from the contour cells are delayed and the excitation wave goes beyond the body of the pin or, on the contrary, the drag wave enters the body of the pin, the process stops after the collisions of these waves end. Then logic is triggered, which classifies the cell as excited or inhibited, depending on most neighbors. Such logic straightens the overlaps described above.

Thus, background damping is performed due to "braking" waves sent from all edges of the image. Simultaneously, excitation waves propagate from a previously determined geometric center. As a result, in the places of collision, the outer contours of the group of parts are formed. Focusing primarily on them, the control system begins the process of understanding the scene. She turns the pin model sequentially in her memory, achieving the same look as in the image. In the geometric center, the apparent size of the cylindrical part is measured. From the similarity of triangles, the distance from the matrix to the geometric center of the pin is determined. Then, by measuring the apparent size of the length of the pin, and calculating the length of the pin located parallel to the horizontal plane at a distance that was determined a little earlier, we can determine the angle of inclination of the axis of the pin to the horizon.

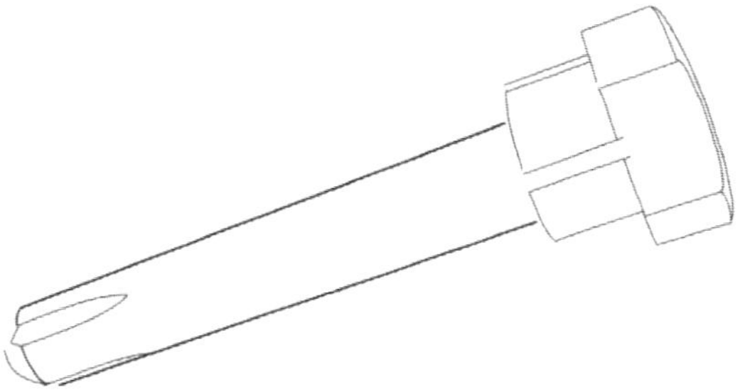

Figure 4. Result of eliminating noise and determining contour lines (rotated 90 degrees).

The information obtained allows the control system of the manipulator to choose the location of the supply of the manipulator - in this case, on the left side, since the pin is in contact with other parts on the right, deploy the grip parallel to the axis of symmetry of the pin and knowing the distance to the pin to capture it.

5. Results and discussion

The values obtained from the recognition matrix are summarized in table 1.

Table 1. Measurement Results

	Diameter mm	*Length mm*	Angle vert. hail	*Length meas. mm*	Diameter meas. mm	*angle horiz. hail*
value	12	92	16.26	125	17	8.1

The control system had to deploy the pin model in memory by 16.26 degrees with respect to the vertical axis of the image and by 8.1 degrees with respect to the image plane, which in turn was removed from the recognition matrix by 30 centimeters. Now the control system is looking for the reaction closest to the given one in the set prepared in advance and starts its implementation.

6. Conclusion

Thus, the study of conclusion mechanisms during hybridization of neural and logical intelligent systems leads to an efficient "island" allocation of sections of neural networks that perform a specific function and the inclusion of logically processing information subsystems between these sections, which can significantly expand the range of managerial tasks to be solved. Such hybridization has significant prospects for practical application.

References

[1] Golovko V A and Krasnoproshin V V 2017 *Neural network data processing technologies* (Minsk: BSU)
[2] Abdrashitov R T and Pishchukhin A M 1990 *Device for modeling the nervous system* (copyright certificate RUS No. 1615755. Application No. 4377111 of 02/10/1988)
[3] Golovko N A 2013 *Automated energy-saving neurocontrol of the temperature regime of the continuous hot dip galvanizing unit* Diss. Cand. tech. Sciences, Magnitogorsk
[4] Kolesnikov A V, Kirikov I A and Listopad S V 2014 *Hybrid Intellectual Systems with Self-Organization: Coordination, Consistency, Dispute* (M.: IPI RAS)
[5] Averkin A N 2016 Hybrid modular neural networks *Economics, Statistics and Informatics* **4** 8-11
[6] LeCun Y, Boser B, Denker Jet al. 1990 Handwritten Digit Recognition with a Back-Propagation Network *Neural Information Processing Systems Conference* **2** 396–404
[7] Rosenblatt F 1965 Principles of neurodynamics: perceptrons and the theory of brain mechanisms (Moscow: MIR)
[8] Le Cun Y et al 1989 Backpropagation applied to handwritten zip code recognition *Neural computation* **1 (4)** 541–551
[9] Le Cun Y et al 1999 *In shape, contour and grouping in computer vision* (B.; Heidelberg) 319-345
[10] Le Cun Y et al 1989 Gradient-based learning applied to document recognition *Proc. of the **IEEE*** **86 (11)** 2278-2324
[11] Batischev V I, Kuzmin M I, Pischukhin A M, Solovyov N A 2016 System of computer vision for cold-rolled metal quality control *International Review of Automatic Control* **9 (4)** 259-263

IOP Conf. Series: Materials Science and Engineering **843** (2020) 012022 doi:10.1088/1757-899X/843/1/012022

Optimisation of metal-cutting tool geometry based on chip formation requirement

M A Korchuganova, Iurii Guskov, V R Ponurovskaya and A Syrbakov

Novosibirsk State Agrarian University, Dobrolyubova str., 160, 630039, Novosibirsk, Russia

E-mail: kma77@list.ru

Abstract. This study of the process of chip formation and curling when cutting ductile materials supplements the work of well-known authors in this field. The relevance of theoretical and experimental research stems from determination of the design parameters of chip formation by the flat spiral shape. The study used tools and materials that closely resemble those used in machine repair shops. A distinctive feature of this study is the use of theoretical and experimental data to determine the conditions of chip breaking and to design chip-forming elements on the cutting face of a metal-cutting tool. The obtained results allowed theoretical calculation of the geometrical parameters of a metal-cutting tool in constrained cutting and determination of the areas of chip formation with a high bulk density.

1. Introduction

Modern machine-building industry has a fairly wide range of metal-cutting tools at its disposal. This is especially true for lathe tools. Tools with replaceable indexable inserts are considered to be the most cost-effective. The main structural elements on the cutting face of a metal-cutting tool designed to form and remove chips from the cutting zone are the width of the chip-breaking step, its height, and cutting edge inclination angle λ, indicated in Fig. 1 as the angle between the Y axis and the inclined cutting edge. The latter angle determines the chip flow direction. The geometrical parameters of these elements are mainly assigned on the basis of empirical studies after conducting the relevant experiments, which is indicated by the relationship observed between the centres of the areas of the recommended cutting conditions in the S/t coordinates, which is equal to 0.1 [2]. The presence of various geometries of the cutting face in replaceable indexable inserts and the constraint S/t = 0.1 may be considered unreasonable. The use of expensive tools in agricultural repair shops is economically impractical. Therefore, there is a need for a simplified pattern of chip formation and curling, which could provide adequate information about the processes occurring during the transformation of the cut layer into chips. These phenomena should be examined in the plane of deformation that passes through the direction of the movement vectors of the cut layer and the chips [1].

The objective of this study is to develop a simplified pattern of chip formation and curling, which could provide adequate information about the processes occurring during the transformation of the cut layer into chips when workpieces are machined during the repair and reconditioning of machine components. Chip formation should be examined in the plane of deformation that passes through the direction of the movement vectors of the cut layer and the chips.

2. Theoretical section

Chip formation in constrained cutting has certain particular features. Here, several straight and curved cutting blades are involved in the cutting process. This results in a variable thickness of the cut layer, sometimes a variable cutting speed, and, as a consequence, a variable rate of deformation of metal particles. Studies of the features of the constrained cutting process largely deal with finding the shape of the conditional shear surface and determining the angle of chip deviation from the plane perpendicular to the main cutting edge [6, 10 - 12].

In constrained cutting patterns, it's not the length that plays a vital role, but the contact spot, which can be obtained as follows [3, 8]: the contact length calculated for the specific cutting layer thickness a_i is marked from the cutting edge in the chip flow direction η. The geometry of this contact for constrained cutting of brittle materials is based on the experimentally observed symmetry of the contact spot shape [11, 12].

It follows from the examined chip formation pattern that the chip flow direction defined by the angle η_c is very important for studying the mechanics of constrained metal cutting. It has been found that the initial angle of the chip flow is determined by the ratio of the projected areas of the conditional shear surface to the coordinate planes F_x and F_y [2]:

$$tg\eta = F_x / F_y \quad (1)$$

Using CAD/CAM/CAE software eliminates the need for numerical and approximate calculations of the projected areas and the positions of the centres of gravity of projections and cross-sections. We used the KOMPAS-3D software in our calculations. Figure 1 shows a version of simulation of the chip element flow over the cutting face of a metal-cutting tool built in the above-mentioned software. KOMPAS-3D can calculate not only the coordinates of the centres of gravity of projections of conditional shear surfaces, but also the values of moments of inertia, which affect the shape of the chips being formed [3, 7].

Chip formation patterns in constrained metal cutting can help determine the direction of chip movement at the beginning of chip formation and explain the complex shape of the cross-section of chips; however, the above patterns cannot help predict the shape and dimensions of the chips being formed along its length. Chip curling patterns solve this problem.

Certain chip formation parameters during oblique constrained cutting are described fairly well [7, 8, 9, 11, 12]. The main disadvantage of the existing models is that they examine the deformation of chips in the plane perpendicular to the supporting surface. We suggest examining a model of chip formation in space. A distinguishing feature of the proposed model is the interconnection between the chip formation and curling zones. The layer of metal being cut is deformed into chips, passing the regions of elastic, plastic flow, elastic recovery, and a rigid one. During continuous chip formation, the chips, moving on the cutting face of the tool at a constant speed, are in a state of dynamic equilibrium. The chip flow direction is in the plane of action of the resultant of forces and the cutting speed. Applying weak external forces to the chips at the time of their formation changes their shape, because the region of plastic flow exists constantly, while the rigid portion of the chips, which becomes free from the external cutting forces, remains connected thereto. In general, chip curling occurs in two planes: the plane of the cutting face of the tool and the plane of the chip flow perpendicular to the cutting face.

In terms of cutting process stability, safety of machining, transport and recycling conditions, the most convenient chips are chips curled in a flat spiral, which have the lowest weight factor for flow chips [3]. This shape comes from chip curling in the cutting plane of their flow.

Figure 1, a shows the cutting portion of the tool with an element of the flowing chips. Let us assume that the normal and tangential loads on the force contact spot between the chips and the cutting face are distributed according to a certain law. As a first approximation, it can be assumed that the stresses on the conditional shear surface and the tangential loads on the contact spot are evenly distributed. If the chips encounter no obstacles along the way, assuming the chip movement speed on the cutting face of the tool to be constant, we find that the forces acting on the chips from the conditional shear plane *ABD* and the cutting face *AA'B'B* are balanced. This will fulfil the condition of circular curling shown in Figure 1, b.

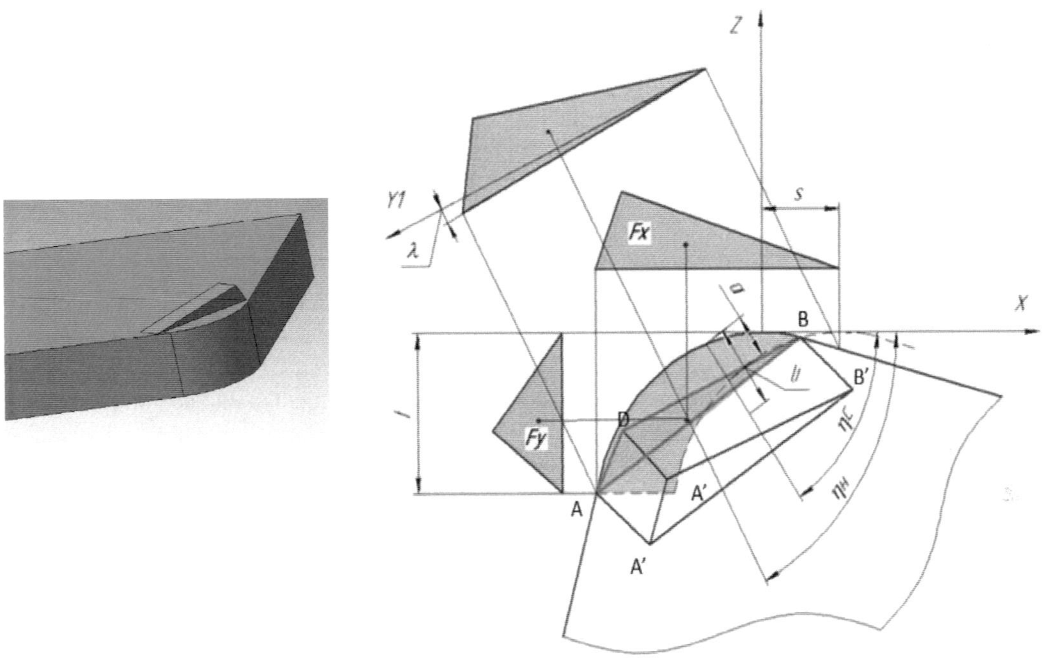

a) b)

Figure 1. Diagram for determining the centres of gravity of the cross-section of chips, the force contact spot, and the chip flow angle, a – diagram of a chip element on the cutting face of a metal-cutting tool; b – diagram of projections of the conditional shear surface.

In general, the points of application of the chip pressure to the tool at the centre of gravity of the conditional shear surface *ABD* and the centre of the contact spot *AA'B'B* do not coincide. This results in a bending moment that defines the so-called "natural" curling of chips of general form. In this case, the arm between the centres of forces, specifically the centres of gravity of the conditional shear surface and the chip contact spot, determines the bend of chips in the vertical plane towards the cutting face [7]. Circular chip curling occurs when the forces of chip formation and friction on the cutting face are in the same vertical plane. This condition is fulfilled when the horizontal coordinate of the centre of gravity of the cross-section of chips coincides with an equivalent coordinate of the centre of gravity of the force contact area between the chips and the cutting face of the tool.

To design the cutting face of a metal-cutting tool that fulfils the conditions for circular chip curling in the cutting plane of chip flow, it is necessary to know the chip flow direction and the contact spot dimensions. Generally, the cutting edge *AB* can be described by the sequence of points (x_i, y_i, z_i), and its previous position *A'B'* can be described by the sequence of points (x_i-s, y_i, z_i), where *s* is the feed rate, mm/rev. Let us examine the cut layer cross-section, which is formed by two consecutive positions of the cutting blade (Fig. 1). In the case of a curved blade of arbitrary shape, for certain cutting depths *t* and feed rates *s*, it is always possible to determine the coordinates of the nodal points of the cut layer cross-section: AA'B. Let us assume that the shear angle β_1 in the chip flow direction for any point of the working section is constant and apply the pattern with a single conditional shear surface. The ratio of the projected areas of the conditional shear surface to the coordinate planes F_x and F_y sets the initial chip flow angle η_n, and the actual chip flow angle η_d depends only on the cutting depth and the feed rate, which determine the nodal points of the cut layer cross-section and the cutting edge geometry in plan view.

Knowing the chip flow angle will help predict the chip movement on the cutting face of a metal-cutting tool and design the geometry of replaceable indexable inserts with predetermined properties with greater precision. The proposed methodology can be used to calculate the simplified geometry of replaceable indexable inserts for mechanical sharpening in shops for repair and reconditioning of machine components.

3. Experiment

The main factor that complicates the study of the chip movement pattern on the cutting face is its sensitivity to external forces, which causes changes in the chip flow direction while the machining conditions remain unchanged. In addition, there are no reliable methods for measuring the chip flow angle [3].

The analysis of literature sources has shown that most definitions of the term “chip flow angle” or “chip deviation angle” refer to machining with a straight cutting edge. These angles are measured from the perpendicular to the cutting edge [11, 12].

In machining with a curved cutting edge, the concept of the perpendicular to the main cutting edge becomes irrelevant for chips in general. In our studies, we assumed that the chip flow angle is measured as the angle in the cutting face plane of the insert between the perpendicular in the line defining the chip cross-section *AB* and the workpiece rotation axis.

Various means and methods are used to measure the chip flow direction. One of the simplest and most effective methods is coating the tool with paint or some other material (the most common being a layer of copper) and observing the trace left by the chips on the cutting face after the cutting process is stopped. In this case, even minor oscillations in the chip movement will distort the trace pattern. In addition, it is believed that during the curvilinear motion of chips the trace left on the cutting face defines only a certain average direction of the chip flow within the contact patch between the chips and the cutter. Another method is based on the immediate stop of the cutting process and the direct measurement of the chip flow direction along the chip root. In this case, it is the immediate stop that is very difficult to carry out. A more reliable method of direct observation is taking photographs of the chips when they move on the cutting face [3]. The series of photographs of the instantaneous positions of the moving chips allow examining the process kinematics. However, this method is labour-intensive.

If we replace the taking of photographs of the chip flow with video recording with further digital representation, the process of data processing becomes faster and allows performing a detailed analysis of the image. Figure 2 shows the digital photographs of the chip flow taken within 10 seconds with one-second intervals. These measurements have shown that the chip flow angle varies from 43° to 57° within 10 seconds, with its arithmetic mean value being 50.4°. The calculated value of 50° was obtained using the same machining conditions. The minimum value of the absolute error of the calculated chip flow angle on the cutting face was 2%, the maximum value was 14%, and the average value was 9.2%. Therefore, later in the study, the chip flow angle was measured using the video recording of the chip flow on the cutting face of replaceable indexable inserts.

In the experiments on measuring the chip flow angle, the cutting conditions were chosen such that “flat” or “cylindrical” spiral chips were formed in most cases.

Figure 3 shows the experimental measurement results for the chip flow angle on a flat cutting face. A cemented carbide tool with replaceable indexable inserts with a flat cutting face was used. The 20G steel was chosen as the material to be machined. A very wide range was adopted for the cutting conditions, with the cutting depth varying in the range of t = 0.3–3 mm and the feed rate S = 0.17–1.2 mm/rev.

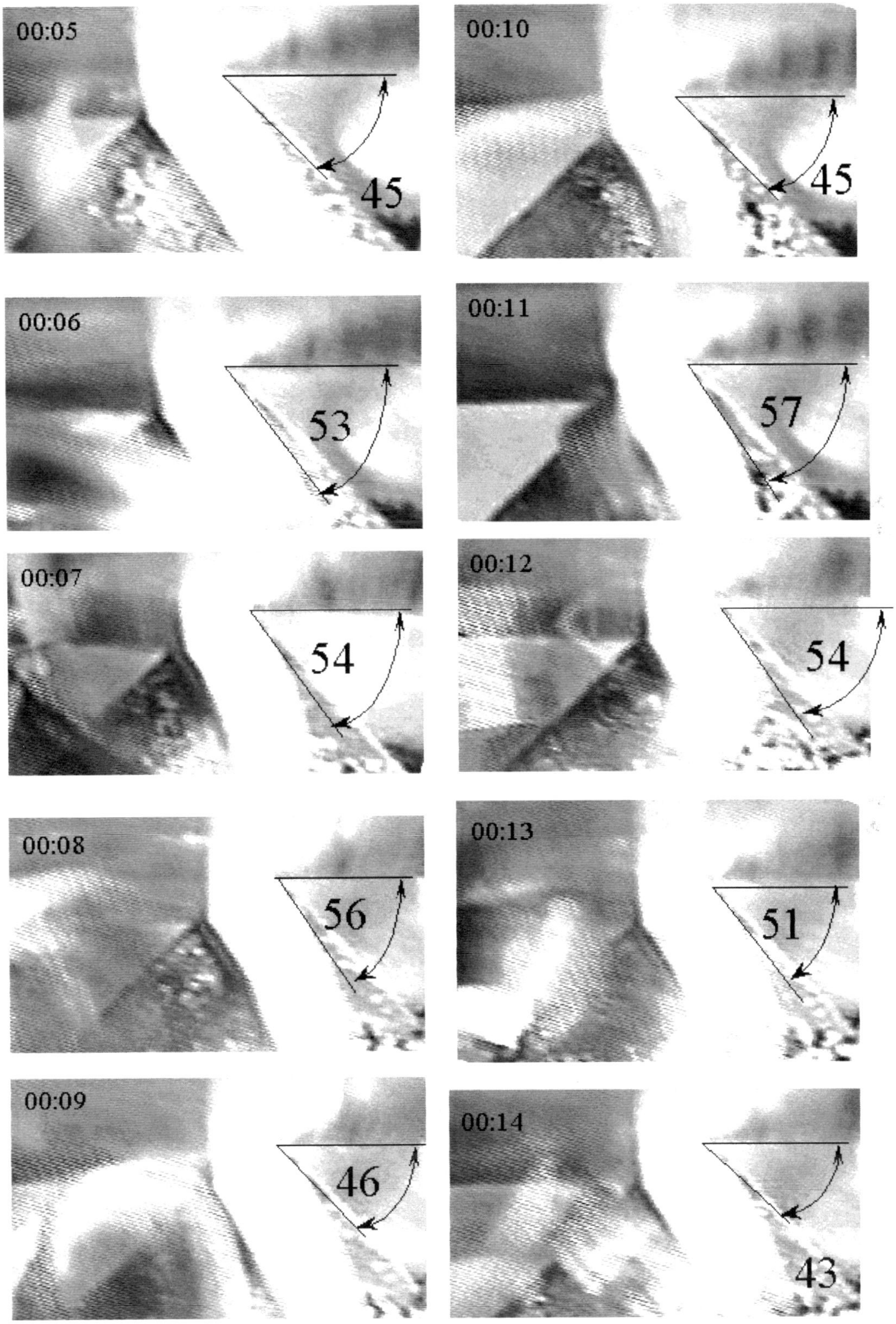

Figure 2. Sequence of images showing the change in the chip flow angle over a 10-second period during formation of helical spiral chips

The cutting speed varied from 120 m/min to 200 m/min. An analysis of the results obtained showed that with an increase in the feed rate, the chip flow angle also increased, while the cutting depth and speed within the specified intervals did not have a significant effect on the chip flow angle. In this respect, we consider the presentation of experimental data depending on the *S/t* relationship to be the most convenient, especially because this coordinate reflects the chip-breaking areas in the manufacturers' catalogues [4, 5]. It follows from this relationship that with an increase in the feed rate the chip flow angle increases, and that this effect is the greatest with comparable values.

A correlation relationship between the chip flow angle and the *S/t* ratio may be constructed (Figure 3). In the case of a linear relationship, the correlation coefficient is 0.93. The analysis of the linear regression equation demonstrated the significance of the coefficients and the equation as a whole.

The theoretical calculations of the chip flow angle and the circular curling conditions and their experimental verification allowed designing replaceable indexable inserts with the required cutting face shape. The complex reticulate surface [2] was replaced with a plane; deviations from the calculated values of the cutting edge inclination angle did not exceed 1.5°.

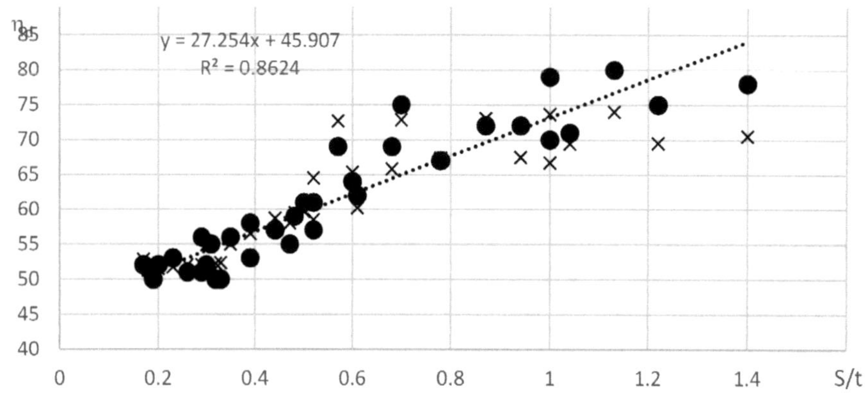

Figure 3. Dependence of the chip flow angle on the cut layer cross-section *S/t*:
• – calculated from the ratio of the projected areas of the conditional shear surface;
x – measured for the conditions: t = 0.3–3 mm and feed rate S = 0.17–1.2 mm/rev, the tip radius was 0.5 mm.

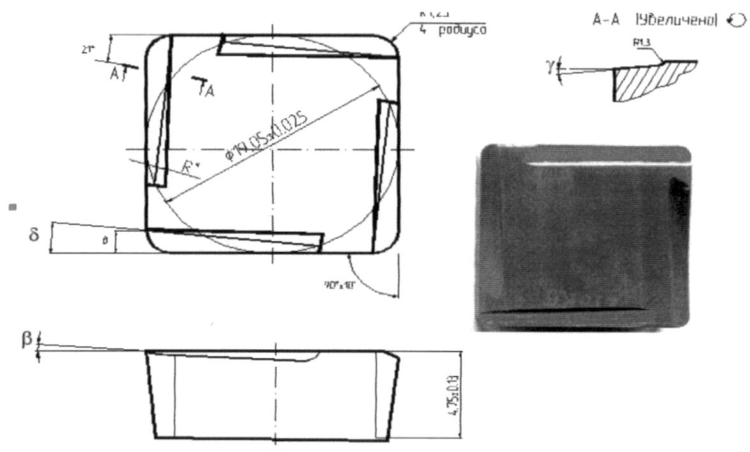

Figure 4. Drawing for sharpening and insert designed from the conditions of circular curling for the following conditions: t = 0.5–2 mm and S = 0.05–0.6 mm/rev

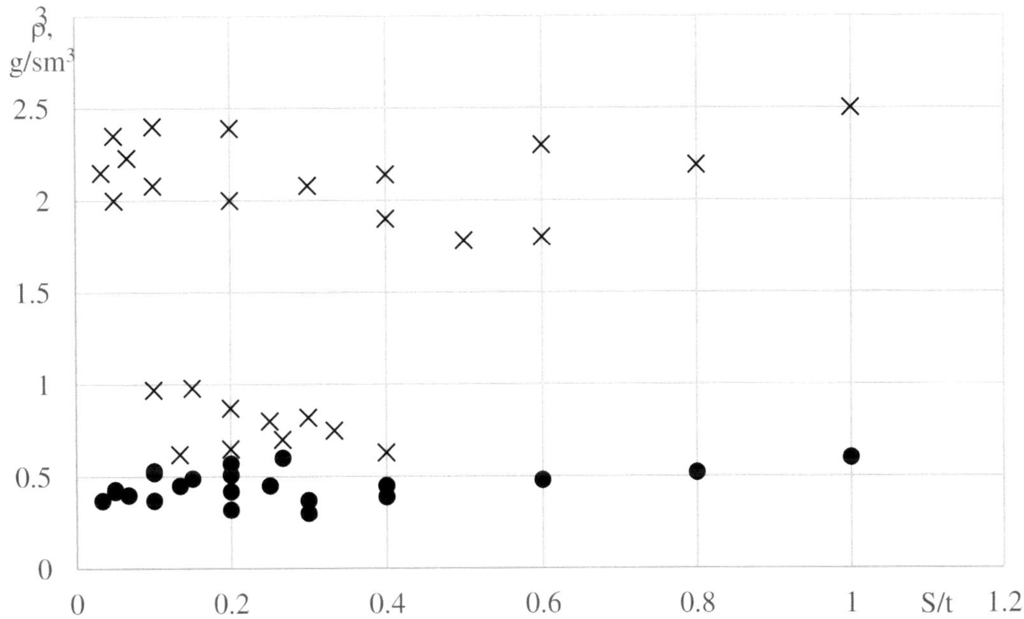

Figure 5. Effect of the S/t value on the bulk density of chips when machining at a cutting depth: x – using an insert with the designed cutting face shape; • – using an insert with a flat cutting face

Experiments were performed to determine the chip-breaking properties of inserts with a flat cutting face and with a radius groove using a CNC lathe, which ensured the required rigidity of the manufacturing system. A cutting tool with the following geometrical parameters was used: $\varphi = 45°$, $\varphi_1 = 45°$, $\alpha = 7°$, $\gamma = 4°$, $\lambda = 3°$. A 20G steel shaft with a diameter of 70 mm was machined.

Two types of inserts were tested in the course of the study: an insert with a flat cutting face, taken as the standard, and an insert with a chip-forming groove. The material used for the insert was T15K6, according to GOST 3882-74. The experimental inserts were produced by modifying the standard insert for the calculated conditions of circular curling.

During the experiments the formed chips were collected and weighed using a precision balance to determine the bulk density. The bulk density ρ was determined by dividing the chip weight by their volume.

During the first stage, the chip-breaking properties of inserts were tested for the design conditions. The tests showed that the standard insert produced flow chips in the form of a helical spiral, whereas the insert with the proposed shape produced finely broken chips with highly stable pieces.

For an exhaustive comparison of the chip-breaking properties of replaceable indexable inserts with the designed shape and those with a flat cutting face, experiments were performed within the cutting depth range t = 0.5–2 mm and the feed rate S = 0.05–0.6 mm/rev (V = 1.67 m/s); the results were used to obtain the chip-breaking diagrams shown in Figure 5.

The results of the study of the chip-breaking properties of replaceable indexable inserts with a flat cutting face showed that in fine-finish and semi-finish cutting conditions, in the depth range t = 0.5–1 mm and S = 0.05–0.6 mm/rev, both flat and helical spiral chips are formed by external longitudinal cutting. The bulk density in this case varies from 1.78 g/cm^3 to 2.5 g/cm^3. When the cutting depth is increased to t = 1.5–2 mm and in the feed rate range S = 0.2–0.6 mm/rev, the shape of the formed chips changes from multi-turn to single-turn and half-turn spirals, which leads to a sharp decrease in the bulk density. Thus, when t = 1 mm and S = 0.2 mm/rev, ρ = 2 g/cm^3, and when t = 2 mm and the feed rate is the same, ρ = 0.97 g/cm^3.

The analysis of the chip shapes produced by cutting with the standard and experimental inserts clearly demonstrates that the proposed shape of a square indexable insert with a chip-forming groove has high chip-breaking properties not only for the design conditions, but also in the expanded range of t and S. This is confirmed by an average 4- to 6-fold reduction in the bulk density of chips.

The economic efficiency achieved as a result of introduction of replaceable indexable inserts with improved chip-breaking properties has several aspects: in the case of mass production, the chip recycling time will be reduced, and for one-off production associated with shops for repair and reconditioning of machine components, the proposed methodology will reduce the expenses for metal-cutting tools and support of the machining process in general.

4. Conclusions

1. The importance of the cutting face shape in chip formation was determined, the shape elements of replaceable indexable inserts that are directly involved in cutting and chip formation processes were identified.
2. The use of the Kompas-3D CAD/CAM/CAE software enabled us to change the method of calculating the circular curling conditions and to simplify the calculation procedure.
3. Based on the theoretical and experimental studies performed, we established that the chip flow angle on a flat cutting face depends only on the cutting edge shape and the cut layer cross-section and, consequently, on (S, t, r, φ, φ_1), but not on the cutting speed. The comparison of the experimental and calculated values of the chip flow angle and the contact length allows using calculated data in the design of the chip-forming elements on the cutting face of replaceable indexable inserts.
4. A methodology was proposed for designing chip-forming elements to form circular chips, the latter being the optimum choice for subsequent recycling.

References

[1] Korchuganova M, Syrbakov A, Chernysheva T, Ivanov G and Gnedasch E 2016 Computer Simulation of Replaceable Many Sider Plates (RMSP) with Enhanced Chip-Breaking Characteristics *IOP Conf. Series: Materials Science and Engineering* **142** 012064 doi:10.1088/1757-899X/142/1/012064

[2] Korchuganova M, Syrbakov A, Chernysheva T, Ivanov G and Korchuganov M 2016 Development of a Metal Cutting Tool Fase in Order to Create the Conditions of Ringed Chips Wrapping *IOP Conf. Series: Materials Science and Engineering* **142** 012063 doi:10.1088/1757-899X/142/1/012063

[3] Korchuganova M A 2012 *Design indexable inserts. Design of cutting tools* (Saarbrucken: LAP LAMBERT Academic Publishing GmbH & Co. KG)

[4] 2016 *Coromant Turning Tools 2012: Catalogue* (Sandviken: AB Sandvik Coromant)

[5] 2016 *Innovations Master Catalog Tooling Sistems 2013* (Kennametall Inc., Latrobe)

[6] Nekrasov R Yu, Starikov A I and Lasukov A A Entering the operative correction machining processes CNC. *IOP Conference Series: Materials Science and Engineering* **91 (1)**

[7] Nakayama K and Ogawa M 1978 Basic rules on the form of chip in metal cutting *CIRP Ann.* **27 (1)** 17 – 21.

[8] Proskokov A V 2013 Theory of Constrained Cutting: Model of Chip Formation with a Developed Plastic-Deformation Zone *Materials Science Forum* **762** 782-789

[9] Proskokov A V and Petrushin S I 2014 Process Chip Formation with a Single Conditional Shear Surface *Applied Mechanics and Materials* **682** 304312

[10] Proskokov A V, Filippov A V and Gorbatenko V V Experimental Study of Plastic Deformation Intensity in Cutting Zone *Applied Mechanics and Materials* **770** 200204

[11] Rosenberg A M and Eremin A N 1956 *Elements of the Theory of the Process of Metal Cutting. Mashgiz* (Moscow)

[12] Grubyy S V 2017 Calculated Parameters of Chip Formation for Restricted Oblique Cutting of Plastic materials *News of higher educational institutions. Engineering* **1(682)** 4-15

IOP Conf. Series: Materials Science and Engineering **843** (2020) 012023 doi:10.1088/1757-899X/843/1/012023

Role of machine building in the development of Russian economy at the present stage

N V Bozo, E V Malysheva and N A Filateva

Novosibirsk State Technical University, 20, K. Marx Ave., Novosibirsk, 630073, Russia

E-mail: bozo_nstu@ngs.ru

Abstract. Machine building is a leading industry in advanced countries. It provides many industries with machines and equipment as it is the main link in the system of intersectoral linkages. Thus, it secures full employment for highly skilled employees, promotes the growth of the level of educational background of the population and contributes to the development of the scientific and technological potential of the country. However, underinvestment in innovation technologies and processes, a limited character of the import substitution model and an insufficient support of the export model greatly hinder rapid and effective development of the machine-building industry. Specific peculiarities of the machine building development such as R&D intensity, labor intensity, and metal intensity, a need in cooperation and consumer orientation influence their geographic location.

1. Introduction

Manufacturing industries in general and machine building play a leading part in industrial production realizing scientific and technological progress, modern methods of organization of production based on the output of priority industries and creating an active part of basic production assets for modern economy. This in its turn affects the major indices of production effectiveness in the economy. Investigation of the importance of the role of engineering was carried out at the country and regional levels.

2. Trends in the development of mechanical engineering in the Russian Federation

The significance of manufacturing industries in the economy of the country depends on a great number of factors which include the level of economic development, demand for products of a certain sector, the degree of outsourcing of particular elements of the value-added chain of manufacturing in service industries and the character of the stimulation policy. The role of natural resources in the economy also influences the development of manufacturing industries. As a result of these factors' effect, the GNP share of manufacturing industries in leading countries varies from 10 % in Great Britain to 33 % in China. In 2018, the share of manufacturing industries in gross value added (GVA) in the Russian Federation amounted to 14 % and in the long term it tends to decrease (Table 1). However, recently positive changes have become evident.

Until 2008, manufacturing industries were actively developing, which was due to highly favorable economic conditions, great internal consumer and investment demand, as well as low real interest rates.

Table 1. Share of machine building and manufacturing industry in GVA, %

	2009	2010	2011	2012	2013	2014	2015	2016	2017	2018
Manufacturing industry	14.8	14.8	13.4	13.6	12.9	13.1	13.8	13.0	13.7	14.3
Machine building	3.2	3.4	3,0	3.3	3.1	3.2	3.6	3.1	3.4	3.2

According to ROSSTAT (Russian statistics service) the manufacturing sector in the Russian Federation remains one of the most significant sectors in terms of the GVA share in GNP and is inferior only to the commerce sector in the number of the employed. [1] An average number of people employed in the manufacturing industry is quite stable and is about 6 % of the total number of people employed in Russia. The number of manufacturing enterprises has considerably decreased (from 402.5 thousand to 309.8 thousand) since 2010 though the share in the total number of enterprises remains at the same level (7.3 %-7.4 %).

Despite some negative tendencies in the manufacturing industry the share of machine building output in the structure of manufacturing industries in the Russian Federation is on average 23.5 % and tends to increase. Nonetheless, the contribution of machine building to GVA is quite stable and amounts to 3.2 %. The number of people employed in the machine building industry has decreased both in absolute and relative terms (by 8 % since 2010) while the number of machine building enterprises has decreases by half (from 81.3 thousand to 40.7 thousand)

In the structure of export in the Russian Federation the share of manufacturing enterprises (as compared to European countries) is not big. The share of machine building output is about 6.5 % and tends to decrease. But in money terms the export of machines, equipment and transport vehicles increases though at a decelerating rate (Figure 1).

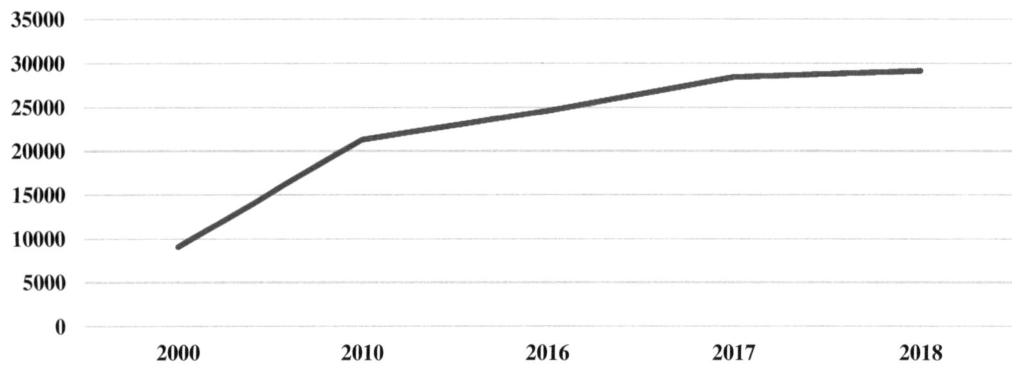

Figure 1. Export of machines, equipment and transport vehicles, million US dollars

The diversification of the industry, training of highly qualified specialists and application of innovation developments, as well as an increasing interest of the government in developing both the machine building industry and other industries having high value added will make it possible for the machine building industry to grow. However, economic slowdown in the world greatly limits positive prospects, which reduces demand for goods.

Analyzing costs for innovations in the manufacturing industry we can observe an insignificant reduction in 2014-2015 after an accelerated growth in 2010-2013. However, costs for innovations continue to grow both in the economy as a whole and in the machine building industry (Figure 2).

We should pay attention to a strong correlation between the amount of costs for innovations and the volume of exports of innovation products. A close correlation between these indices is observed in the manufacturing industry as a whole (0.96 and 0.91 correspondingly). The output volume and exports of manufacturing innovation products reveals a strong linear dependence (r=0.98). All theses indices are less correlated in the machine building industry though the correlation is sufficiently strong.

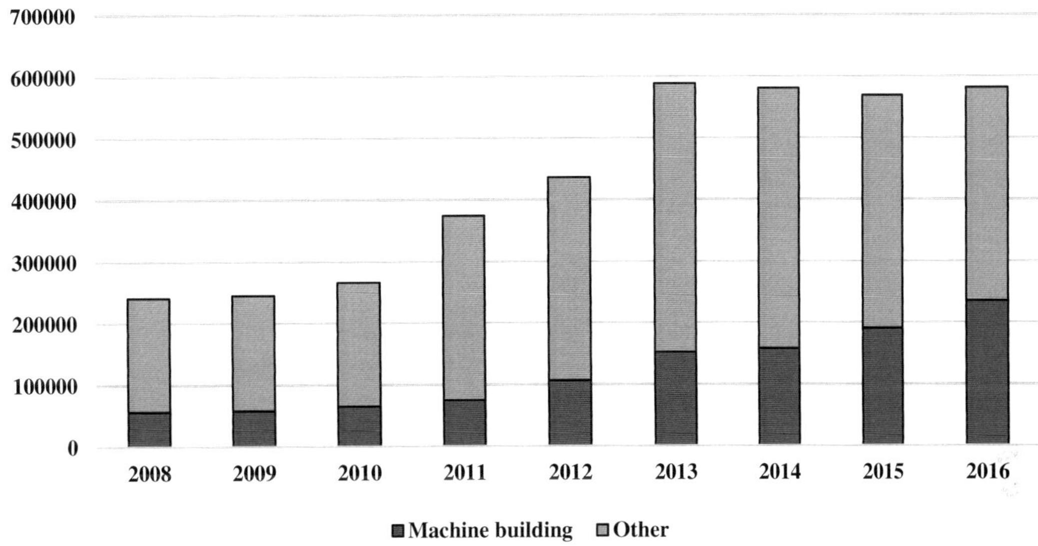

Figure 2. Structure of costs for innovations in the manufacturing industry, million Rub

3. Innovative activity in manufacturing industry

Russian Statistics attributes the manufacture of computers and electronic and optical devices to high-tech manufacturing. The manufacture of electrical equipment, machine-tools for material processing, special purpose machines, transport vehicles, trailers, and semitrailers, as well as other transportation means and equipment is attributed to medium-tech manufacturing of a high level [2].

Innovation activity in Russian machine building was analyzed:

1) by basic high-tech and medium-tech manufacturing of a high level (with regard for changes in OKVED);
2) at the macro level (because of the absence of official innovation industry statistics from RF regions);
3) over the period from 2008 to 2017 (with regard for changes in OKVED; 2017 was analyzed separately);
4) by the structure of costs (marketing, organizational and technological) [3].

The analysis of statistical data on innovation activity of individual types of machinery production has shown:

1) over the period from 2008 to 2016 against the background of decreasing the overall level of innovation activity in manufacturing industries as a whole (0.2 p.p.) the greatest decrease in activity can be seen in the manufacture of aircraft (6.4 p.p.) though the greatest number of these organizations are involved in innovation activity (32.7 % in 2016). In 2017, the highest level of innovation activity was observed in the manufacture of computers and electronic and optical devices (Figure 3). The number of organizations applying organizational innovations decreased most of all.

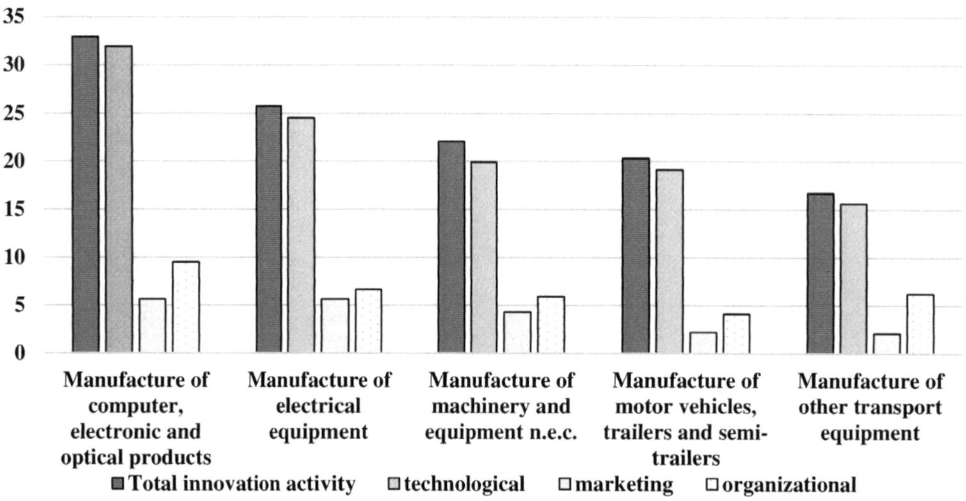

Figure 3. Innovation activity of RF machine building organizations, 2017, %

2) over the period under study a reduction in the innovation cost share (10.9 p.p.) is observed in the manufacturing industry. High-tech machine building was augmenting process innovations against the background of an overwhelming role of technological innovations (98.3 % in 2017). The shares of marketing and organizational innovations are extremely low and are decreasing.
3) The amount of innovation goods, developments and services in the total volume of shipped goods is increasing both in manufacturing industries (by 2 p.p. as of 2017) and in machine building production under analysis. In this case 10.5 % of the innovation products are created by high-tech manufacturing enterprises under government and municipal contracts (Figure 4). The share of exported (high-tech) innovation goods is growing.

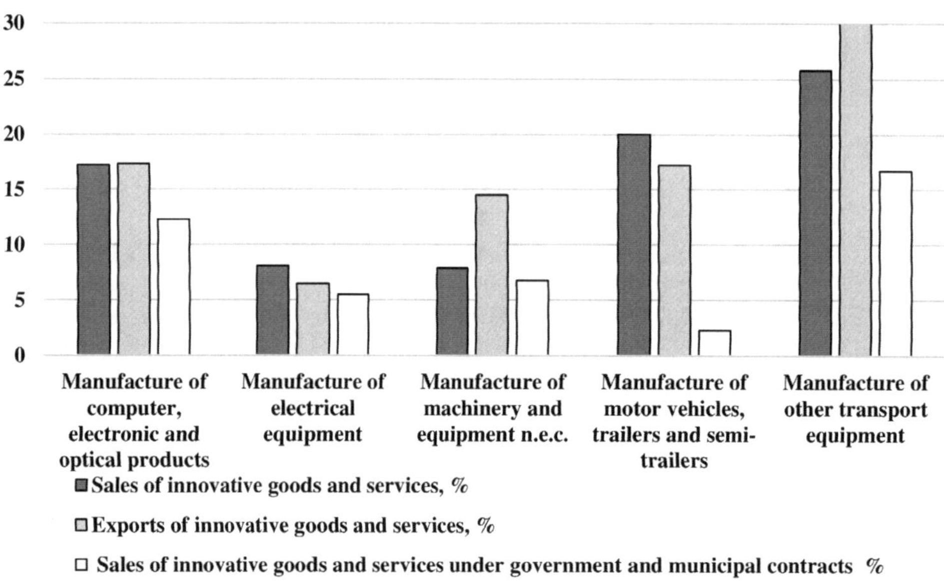

Figure 4. Production of innovation goods, developments and services in RF machine building, 2017, %

4. Analysis of the regional specialization of the Russian industry

The geography of the machine building industry distribution over the Russian Federation territory is quite peculiar. It is assumed that it is undoubtedly influenced by labor intensity, metal intensity and research intensity. Taking into account the fact that the manufacture of machines requires much work time and highly qualified human resources, this industry is often located in densely populated regions.

When we speak of labor intensity, there is no doubt that highly qualified engineering personnel welcome the development and application of new samples of technical facilities. The most famous are such sub-sectors as machine tool manufacture in Moscow, instrument making and production of electronic facilities in Ulyanovsk and Novosibirsk, the aircraft industry in Samara and Kazan. It is an open secret that metal intensity is an inherent factor of the machine building complex. It is this industry that consumes the greatest amounts of ferrous and non-ferrous metals. Thus, it is quite natural that machine building enterprises tend to be located near metallurgical bases. Hence, big heavy machine-building plants are located in the Urals (Yekaterinburg). The focus on the manufacture of state-of-the-art and advanced technical equipment makes it necessary to concentrate on regions with highly developed scientific bases, for example, research institutes and design bureaus (in Moscow, St. Petersburg and Novosibirsk), as well as pilot-producing plants. Focus on the research potential is one of the factors that influence the choice of machine building enterprise locations.

However it is also necessary to mention the orientation of the produced output to the consumer sector. This factor is especially significant for manufacturers producing bulky machines because of the difficulties caused by their transportation. For example, grain-harvesting machines are manufactured in the North Caucasus right in the areas f their use (Taganrog and Rostov-on-Don). Some sectors of the machine building complex are located in regions suitable for cooperation, first of all in terms of their geographical location. Machine building plants are usually located near large transport hubs and highways

The location of the machine building industry on the RF territory makes us ponder over the question which exactly regions can be considered to specialize in machine building. The identification of the current professionalization of the region and the legitimacy of its effective specialization are the significant issues in the system of regional relations. When determining the industries of regional specialization, we should use a system of indicators that has to be confirmed theoretically. These indicators should be related to each other via another geographical division of labor indicators. The geographical division of public labor forms the foundation of regional markets specialization. Thus, the determination of the region participation in social division of labor should be the basis for the identification of professional branches.

The analysis of revealing an industry of region specialization was based on a comprehensive approach with the calculation of such well-known indicators as the localization coefficient, a specialization level, inter regional marketability, a level of industry development in the region, per capita production, etc. All these indicators represent theoretical basics that serve as tools for calculating a quantitative level of the region specialization.

The research is based on the calculation of the specialization ratio which helps to determine a specialization industry of regions, federal cities, and federal territories. This indicator is calculated by using the gross regional product (GRP) which in its turn is one of the most important indicators of the results of economic activity of the region

The ratio of the specialization level ($К_c$) [4] shows the relation of the share of a region' industry in the country's output volume to the share of a region's economy in the economy of the country.

$$К_c = \frac{O_p}{O_c} \times 100 / \frac{X_p}{X_c} \times 100, \tag{1}$$

where O_p is a region industry;
X_p is the region economy;
O_c is a country industry;
X_c is the country economy.

An industry is a specialization industry when specialization ratios are greater than 1.

Specialization ratios for all Federal districts and 85 subjects of the Russian Federation were calculated in the research over the period from 2007 to 2017. The calculations were carried out for all types of economic activity according to the OKVED classifier. The results of the specialization ratio calculation in section D "Manufacturing Industries" for RF federal districts showed that according to statistical data this specialization industry is peculiar to the Central Federal District (CFD), the Northwestern Federal District (NWFD), the Volga Federal District (VFD), and the Siberian Federal District (SFD).

Further more detailed analysis shows that the manufacturing industry is present mostly in the Sverdlovsk and Chelyabinsk regions in the Ural Federal District (UFD). These two regions are the largest metallurgical centers in Russia. Ironworks are concentrated in the Sverdlovsk region and machine building and chemical industries are also highly developed in this region. About 40 % of finished metallurgical items of the total output of the country are produced in the Chelyabinsk region.

Most regions in the Southern Federal District (SFG) specialize in agricultural production. Their natural and climatic conditions as well as the natural and resource potential are the main factors of developing this industry in southern regions of Russia. Agriculture and the hotel and catering industry are the prevailing industries in the North-Caucasian Federal District (NCFD). The fisheries industry forms the basis of the economy of the Far Eastern Federal District (FEFD).

The calculations made revealed specialization industries in each region. The RF subjects whose specialization industry is manufacturing in particular machine building are presented in Table 2. It is these 19 regions that have not changed their specialization industries over the period from 2008 to 2017.

Table 2. Calculated values of specialization ratios for the RF subjects in section D –"Manufacturing Industries" over the period from 2008 to 2017

FD	Subject of the Russian Federation	2008	2009	2010	2011	2012	2013	2014	2015	2016	2017
CFD	Vladimirskaya oblast'	1.77	1.76	1.78	1.71	1.76	1.77	1.83	2.01	2.1	2.72
	Kaluzhskaya oblast'	1.66	1.72	1.89	2.14	2.3	2.08	2.06	1.87	2.15	2.42
	Lipetskaya oblast'	2.78	2.5	2.32	2.19	1.87	1.95	2.49	2.38	2.38	2.73
	Ryazanskaya oblast'	1.32	1.45	1.56	1.66	1.48	1.52	1.62	1.7	1.72	2.1
	Tul'skaya oblast'	1.85	1.61	1.71	2.1	1.92	1.86	2.29	2.3	2.35	2.85
	Yaroslavskaya oblast'	1.4	1.47	1.41	1.53	1.53	1.44	1.48	1.59	1.56	1.91
NWFD	Vologodskaya oblast'	2.7	2.28	2.08	2.16	1.92	2.08	1.83	2.26	2.09	2.21
	Leningradskaya oblast'	1.36	1.58	1.4	1.39	1.3	1.48	1.69	1.68	1.66	1.94
	Novgorodskaya oblast'	1.71	1.78	1.74	1.82	1.96	1.98	2.01	2.07	1.99	2.3
VFD	Respublika Bashkortostan	1.48	1.28	1.58	1.91	2.15	1.74	1.75	1.71	1.61	2.24
	Permskiy kray	1.81	1.57	1.7	1.83	1.79	1.78	1.86	1.79	1.83	2.16
	Kirovskaya oblast'	1.37	1.19	1.29	1.51	1.48	1.42	1.58	1.66	1.66	2.03
	Nizhegorodskaya oblast'	1.65	1.56	1.74	1.7	1.74	1.71	1.77	1.8	1.82	2.06
	Samarskaya oblast'	1.34	1.2	1.39	1.43	1.46	1.43	1.46	1.39	1.3	1.55
	Ul'yanovskaya oblast'	1.11	1.03	1.18	1.23	1.28	1.26	1.47	1.5	1.55	1.88
UFD	Sverdlovskaya oblast'	1.71	1.63	1.65	1.57	1.56	1.54	1.65	1.77	1.79	1.92
	Chelyabinskaya oblast'	2.09	2.04	1.98	2.05	2.04	2.07	1.8	2.05	2.09	2.05
SFD	Krasnoyarskiy kray	1.95	2.02	1.96	1.92	1.72	1.63	1.89	1.88	1.88	1.94
	Omskaya oblast'	2.02	2.11	1.97	2.1	2.18	2.11	2.29	2.1	2.18	2.7

*FD – Federal District

IOP Conf. Series: Materials Science and Engineering **843** (2020) 012023 doi:10.1088/1757-899X/843/1/012023

After analyzing all regions and their industries it can be concluded that regions with a great resource-based potential are least diversified. Regions with high social and economic indicators not involved in the mining industry changed their specialization several times over the period under study. It is caused by the fact that a great number of industries are developed in these regions.

Regions specialized in machine building faced a number of obvious problems and managed to solve them. First of all, it is necessary to change the specialization of the industry which for a long time used to be a defense industry. A major part of smaller human settlements failed to cope with diversification on their own. And without financial support it will be impossible to do after a long time. We can add here rare opportunities of upgrading equipment and as a result obvious machine idly time and output of products whose quality does not conform to international standards.

One of the major tasks of the machine building industry is radical reconstruction and advanced growth of such industries as machine-tool construction, instrument-making, electrical and electronic industry, and manufacture of computers, which will allow Russia to gather pace to approach the global economic level

5. Conclusion

Russian machine building is characterized by a whole range of peculiarities:

1) Machine building is a leading industry of the Russian economy dominating in the manufacturing industry by all indicators. Despite the reduction in the number of enterprises and people employed in the industry, the contribution of machine building to GDP of Russia remains stable.
2) Investments in innovations in the manufacturing industry are practically at the same level. However, costs for innovations in machine building increase, which promotes the growth of production and exports of innovation products, developments and services.
3) Against the background of decreasing innovation activity of machine building enterprises, an extremely low share of costs for measures related to a fuller satisfaction of needs, development of new markets, mastering of new methods of running business, organization of production, and establishment of foreign relations is observed.
4) The share of innovation products, developments and services including exported products grows with the support of the government.
5) Not all well-known enterprises in the Russian machine building industry are located in the regions whose specialization industry is manufacturing.

The government policy should take into account the above peculiarities when developing measures to stimulate the development of the machine building industry. Innovative and technological renewal and upgrade of the real sector industries and the modernization of the economy create favorable conditions for an intensive growth of production volumes in the machine building industry. Investment support of innovation activity in these industries has a multiplicative effect in the related industries and first of all in metallurgy. But it only the continuity of the process of modernization of the fixed capital in the consumer-oriented industries with a constant flow of facilities from the machine building industry will provide the sustainability of the process of modernization of the economy as a whole.

References

[1] Rybalka A I 2018 *Risk factors of the industries of the processing industry* ENSR **03**
[2] Litvintseva G P and Gakhova N A 2016 *Actual problems of electronic instrument engineering (APEIE–2016)* **1** 256–262
[3] Gokhberg L, Ditkovskiy K, Kuznetsova I 2019 *Indicators of Innovation in the Russian Federation* Data Book
[4] Bozo N V 2012 *Regional economy*

IOP Conf. Series: Materials Science and Engineering **843** (2020) 012024 doi:10.1088/1757-899X/843/1/012024

Applications of Discrete Element Method (DEM) in modeling the impact of dynamic and technological parameters on the material movement on the vibrating screen surface

Nguyen Van Xo[1] and Nguyen Khac Linh[2]

[1] Ha Noi University of Mining and Geology, 18, Pho Vien Street, Duc Thang Ward, Bac Tu Liem District, Ha Noi, 90000 Vietnam
[2] Saint-Petersburg Mining University, 2, 21st Line, St Petersburg 199106, Russia

E-mail: khaclinhhumg@gmail.com

Abstract. A vibrating screen is a device for sorting materials, which is widely used in screening plants [1-3]. The availability of a vibrating screen relies on various parameters, such as dynamic parameters, technological parameters, and so forth [4-8]. When using the traditional method to calculate the design and determine the reasonable parameters for the machine to work efficiently, it would be rather challenging, and the accuracy is high. Also, this process would likely take a lot of time, as well as an increasing cost. This study aims to apply the Discrete Element Method (DEM) to simulate the influence of dynamic parameters and technological parameters on the operating capability of the machine. The results of the paper are graphs showing the importance of the oscillation frequency, the amplitude, the incline angle of the screening surface, the angle of oscillation on material velocity, and material elevation on the sieve. These are essential parameters that significantly affect the performance and productivity of the screen. The findings of the study may be useful for scientists to design parameters for vibrating screen reliability.

1. Introduction

A vibrating screen is a sorting machine, used to split bulk materials with various kinds into different size particles thanks to rails. The vibrating screen working principle is given in Figure 1. When the vibrator assembly (exciter) (1) works, the two the eccentric shafts which have counterweights will rotate simultaneously in opposite directions, creating centrifugal forces. The force components acting along x-x axis will be opposite and cancel out, while the force components acting along y-y axis will combine to produce oscillating forces. This force is transferred to the sieve boat (3), associated with springs (2) generating vibrations and carrying out the sieving process to classify materials [9].

When the vibrating screen operates, it depends on the dynamics and technology of the sieve, such as the vibration frequency f, the vibration amplitude A, the angle of vibration direction of screen δ, and the inclined angle of screen α_o. Hence, the material on the vibrating screen may have four relative motions to the screening surface, namely slide-forward motion, slide-back motion, stop (lie on the screening surface) and jump movement. Unless the third scenario in which the vibrating screen does not perform the sieving process, the others operate the screening process [10]. However, since the material travels and jumps up, the screen will be the most effective when working. Therefore, as designing a screen, designers tend to choose this working state.

Nevertheless, it is challenging to employ traditional methods to design parameters for a well-screened sieve. In this study, the author is going to utilize the discrete element method (DEM) to simulate and solve the mentioned problem.

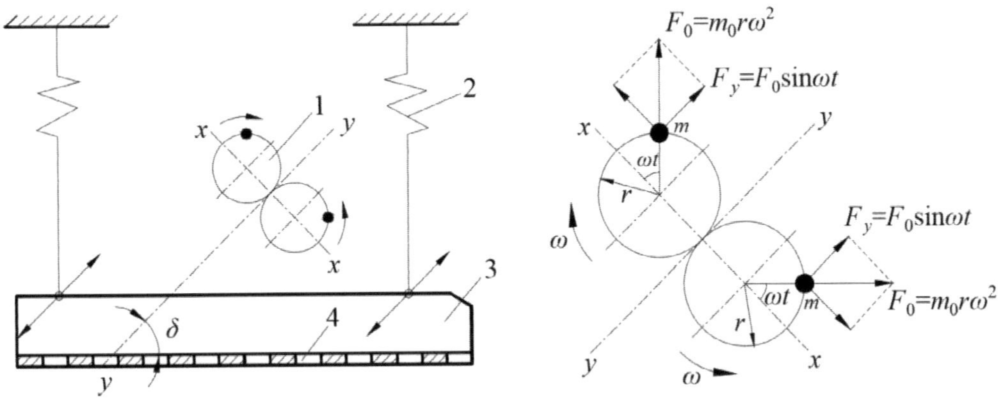

Figure 1: The working principle of a vibrating screen: 1- Unbalance exciter; 2- Springs; 3- Screen body; 4- Screen mesh;
m = mass of each eccentric plate (kg); m_o = total unbalanced mass (kg); t = time (s); r = eccentric radius (mm); ω = angular speed (rad/s); δ = angle of oscillation (°); F_o = exciter amplitude of fluctuations caused by unbalanced mass (N); F_y = excitation force in the y-direction at time t (N).

$$F_y = \sum m_o r\omega^2 \sin\omega t = mr\omega^2 \sin\omega t = F_o \sin\omega t \quad (1)$$

2. Movement principle of material on the screening surface

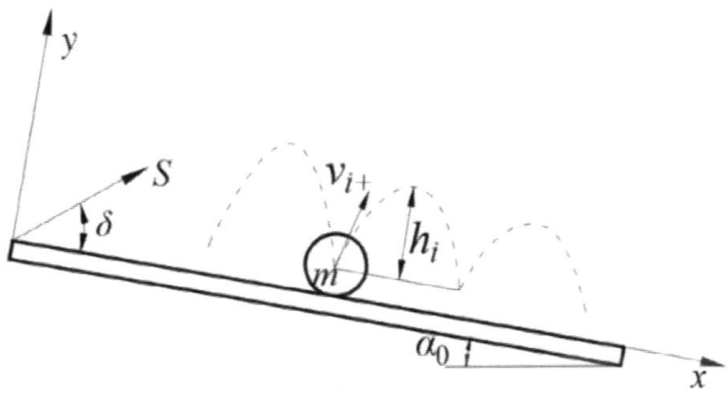

Figure 2: Principle of material movement on the screening surface

The principle of material movement on the screening surface is shown in Figure 2. When the screen is working, the movement position of the screen is determined as follows [11,12]:

$$S = A\sin\omega t \quad (2)$$

Assume that the material falls downwards at the beginning is due to its weight, followed by a collision with the screening surface and bounces.

Consider the collision time between the screening surface and the material at the i th time is t_i. Ignore changes of the speed of the screening surface before and after the collision, as well as the duration of the collision process. By applying the law of momentum conservation, the velocity of the material after the collision in the y-direction is defined as:

$$v_{t_{i+}} = \left|v_{t_{i-}} - \dot{S}_{ti} \sin\delta\right| e \quad (3)$$

Where,

v_{t_i} = the velocity of the material before the ith collision in the y-direction (m/s),
$\dot{S}_{ti}$ = the instantaneous velocity of the sieve at the ith collision (m/s),
e = the elastic coefficient of the material.

Applying the law of conservation of energy, we yield the jump height of the particles compared to the screening surface after the collision as follows:

$$h_i = \frac{v_{t_{i+}}^2}{2g\cos\alpha_o} \quad (4)$$

Where,

g = the gravitational acceleration (m/s^2),
α_o = the inclined angle of screen (^{o}).

Then the average jump height according to the theory of particles is:

$$\bar{h} = \frac{\sum_{i=1}^{n} h_i}{n} \quad (5)$$

Where n = number of bounces of material particles.
Because the material hasn't acted along the x-direction of the screening surface, the material's average velocity, then, will be theoretically calculated using the following equation [16].

$$v_d = \omega A \cos\delta \frac{\pi i_{K_v}^2}{K_v}(1 + \tan\alpha_o \tan\delta) \quad (6)$$

Where,

i_{K_v} = the jump factor (which is the ratio between a jump time and an oscillation period);
K_v = the jumping index, is given as:

$$K_v = \frac{(2\pi f)^2 A \sin\delta}{g\cos\alpha_o} \quad (7)$$

3. Applications of Discrete Element Method (DEM) in modeling the impact of dynamic and technological parameters on the material movement of the vibrating screen surface

The discrete element method (DEM) is a type of numerical method developed in the 1970s and used in a variety of fields (geological engineering, mining techniques, mineral sorting, etc.). The application of DEM can assist us in solving the problem of discrete materials, such as the position of space, the velocity, the force and energy change of particles, and so forth [13-15].

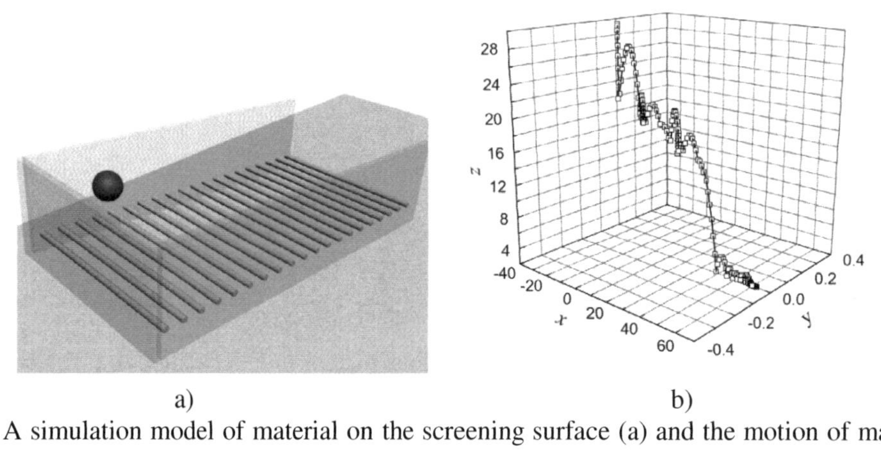

a) b)

Figure 3: A simulation model of material on the screening surface (a) and the motion of materials in space (b)

In this article, the author uses DEM to build up a simulation model of the material on the sieve (Figure 3a) with the following basic parameters, including the size of the screen 60x30 cm, the screen mesh a = 2 cm, the diameter of material d = 4 cm, the elastic coefficient of material e = 0.5. After the simulation, achieving the result of the motion of materials in space (Figure 3b) and some results, as shown in Figures 4 to 7.

- ***Effect of the vibration frequency f:***

If the vibration amplitude, the inclined angle of the screen and the angle of vibration direction of the screen are constant (i.e., A = 5.5 mm, $\alpha_o = 6^0$, δ = 50^0), and only adjust the vibration frequency of the screen. The trajectory of material is drawn in Figure 4.

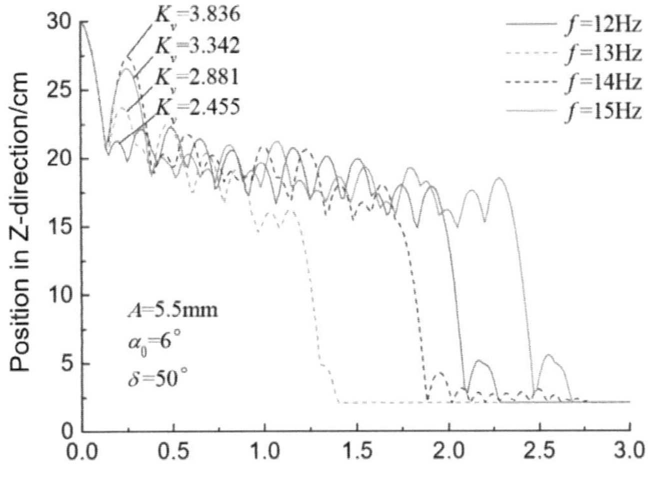

Figure 4: Effect of the vibration frequency on material movement

Figure 4 illustrates that when the vibration frequency of the screen increases, the average speed of the material increases initially and then decreases. The reason is that when the vibration frequency increases, the number of material collisions with the surface of the screen increases and occurs other random collisions, even happen the phenomenon of "jumps later", thereby increasing the number of jumps of material on the sieve and extending the sieving time. When f = 13 Hz, the number of jumps on the surface of the screen equals 6, the time to complete the sieving process is 1.15 s, the average speed of the material yields the maximum value. By contrast, when f = 15 Hz, the number of material jumps on the screening surface is 16, the completion time of the sieving process is 2,267s, and the average speed, as well as the average jump height of the material, hit the minimum values.

From the simulation results (Figure 4), we find out the average velocity and the average jump height of the material (v, h). Also, as compared with the theoretically calculated value (v_d, $\bar{h}$), we have the data, as shown in table 1.

Table 1. The value of average velocity and average jump height when adjusting the vibration frequency

f (Hz)	12	13	14	15
v (m/s)	0.2978	0.4458	0.2140	0.2056
v_d (m/s)	0.2849	0.3087	0.3324	0.3562
h (cm)	26.4468	25.1961	26.4119	22.6399
$\bar{h}$ (cm)	23.4401	18.6786	21.9001	17.5194

Table 1 reveals that when the vibration frequency changes, the average velocity and average jump height of the material change slightly, which means that the frequency of vibration only has a small effect on the average velocity and the average jump height of the material. On the other hand, when f

= 13 Hz, the average velocity and the average jump height of the material are both relatively large values.

- ***Effect of the vibration amplitude A:***

If the frequency of vibration, the inclined angle of the screen and the angle of vibration direction of the screen are unchanged (i.e., f = 14 Hz, $\alpha_o = 6^0$, δ = 50^0), only vary the vibration amplitude, the trajectory of material is shown in Figure 5.

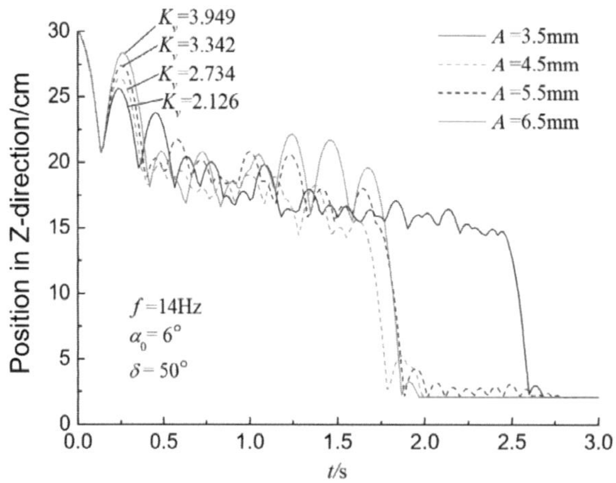

Figure 5: Effect of the vibration amplitude on material movement

From Figure 5, it is evident that as the vibration amplitude of the screen increases, the relative collision velocity between the material and the screening surface increases, the average speed and the average jump height of the material increase, the number of jumps of material decreases (similar to the theory). When A = 3.5 mm, the number of jumps of material on the screening surface is 20 times, the completion time of the screening process is 2,417s, the average speed and the average jump height reach the minimum value.

When increasing the value of A, the required time for accomplishing the sieving process decreases (approximately 1,567 s), and if A = 6.5 mm, the average velocity and the average jump height of the material hit the maximum value.

From the simulation results (Figure 5), the average velocity and the average jump height of material (v, h) are computed, and as compared with the theoretically calculated value (v_d, $\bar{h}$), we yield the data, as shown in table 2.

Table 2. The value of average velocity and average jump height when the vibration amplitude changes

A (mm)	3.5	4.5	5.5	6.5
v (m/s)	0.2078	0.2088	0.2140	0.2977
v_d (m/s)	0.2116	0.2417	0.3324	0.3929
h (cm)	19.4694	21.5484	26.4119	40.9729
$\bar{h}$ (cm)	19.7285	19.8357	21.9001	24.2023

From Table 2, one thing that can be described is the vibration amplitude of the screen shifts; the average velocity of the material also changes relatively. But the average jump height of the material varies greatly, which means the amplitude of the sieve relatively fluctuates to the average velocity of the material. Besides, this has a considerable influence on the average jump height of the material. Thus, paying attention is required during screen design processes. The selection of reasonable parameters means that the vibration amplitude only affects relative to the average velocity of the material, but has a significant influence on the average jump height of the material. As a result, when designing sieves, we need to choose proper parameters (In terms of materials which are difficult to screen, we should opt-out such a large amplitude when the screening machine is likely to achieve high velocity, high jump, and high efficiency).

- ***Effect of the inclined angle of screening surface α_o***

If the vibration frequency, the vibration amplitude and the angle of vibration direction of screen are constant (i.e., f = 14 Hz, A = 5.5 mm, $\delta = 50^0$), the inclined angle of screening surface is altered, the trajectory of material is mapped in Figure 6.

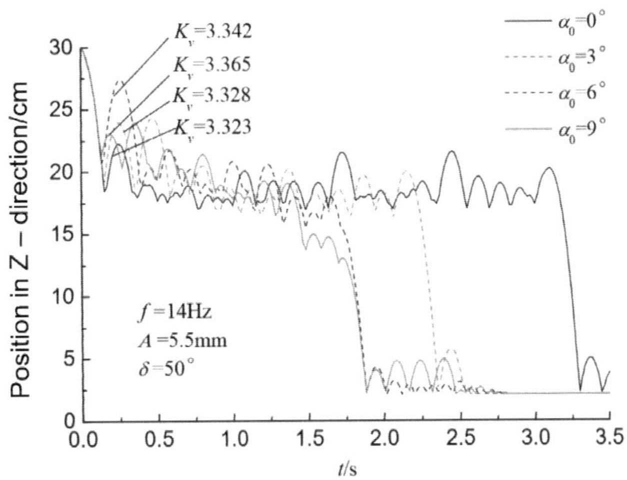

Figure 6. Effect of the inclined angle of the screen on material movement

The implication of Figure 6 is that when the inclined angle of the screening surface increases, the number of jumps of the material will decrease, the average velocity will increase, and the average jump height will decrease. When $\alpha_0 = 0^0$, then the number of jumps of the material is 20 times, the time to complete the sieving process is maximum 3.3 s, the smallest average velocity. When $\alpha_0 = 6^0$, the average jump height of the material reaches the maximum value. When $\alpha_0 = 9^0$, the initial distance between the surface screen and the material is small, the relative speed between screen and material reaches the minimum value, resulting in low jump height.

From the simulation results (Figure 6), we find the average velocity and the average jump height of material (v, h) and as compared with the theoretically calculated value (v_d, $\bar{h}$), we yield the numbers, as shown in table 3.

Table 3. The value of average velocity and average jump height when the inclined angle of screen changes

α_o (°)	0	3	6	9
v (m/s)	0.2477	0.2917	0.2509	0.3169
v_d (m/s)	0.2954	0.3139	0.3324	0.3512
h (cm)	21.5082	22.6978	26.4119	15.1523
$\bar{h}$ (cm)	18.4471	16.7279	21.9001	15.3953

Results are given in Table 3 shows that when the inclined angle of the screen changes, the average velocity and average jump height of the material change slightly. To put it simply, the angle of the screening surface has little effect on the average velocity and the average jump height of the material. When the value of the inclined angle of the screening surface lies within the range of 3^0 - 6^0, the average velocity and average jump height will reach the maximum value.

- ***Effect of the angle of vibration direction of screen δ***

If the vibration frequency, the vibration amplitude and the inclined angle of screening surface are constant (i.e., f = 14 Hz, A = 5.5 mm, α_0 = 60), only changes the angle of vibration direction of the screen; the trajectory of material is represented in Figure 7.

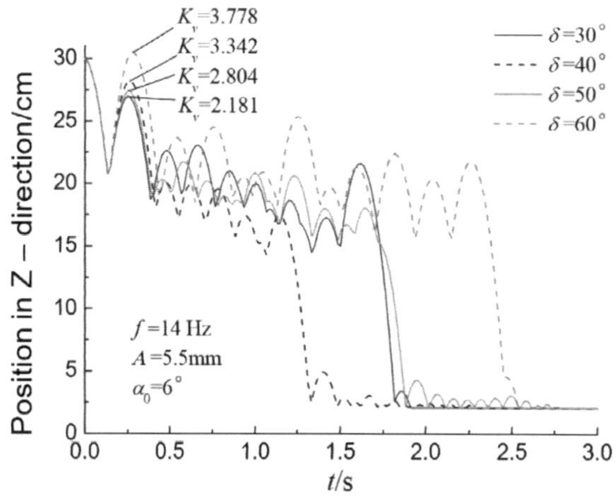

Figure 7. Effect of the angle of vibration direction of the screen on material movement

As demonstrated in Figure 7, when the angle of the vibration direction of the screen increases, the average jump height increases, but the average velocity decreases. When δ=40^0, the number of jumps of the material equals 6, the average speed reaches the maximum value, the shortest time to complete the screen is 1,067s. When δ=60^0, the number of jumps of the material is 11, the average velocity reaches the smallest value, the time to complete the most prolonged sieving process (2.15s), and the average jump height hits the maximum value.

From the simulation results (Figure 7), it is definite that the average velocity and the average jump height of material (v, h), and as compared with the theoretically calculated value (v_d, $\bar{h}$), computed numbers are listed in table 4.

Table 4. The value of average velocity and average jump height when the angle of vibration direction of screen changes

δ (°)	30	40	50	60
v (m/s)	0.3672	0.5293	0.2509	0.1903
v_d (m/s)	0.4222	0.3831	0.3324	0.2716
h (cm)	19.8286	22.5425	26.4119	48.2712
$\bar{h}$ (cm)	15.9063	18.4471	21.9001	25.9641

The above table indicates that when the angle of vibration direction of screen changes, the average velocity and the average jump height of the material change substantially. In other words, the angle of the vibration direction of the screen will dramatically affect the average velocity and the average jump height of the material. Thus, in order to gain great average speed and a high jump at the same time, it is highly recommended to choose the angle of oscillation within the range of about 40^0.

4. Conclusions

By dint of using the discrete element method, the paper has simulated and analyzed in detail the effects of technological parameters, dynamics parameters on the velocity, and jump height of materials on the screening surface when the machine is working. The research results indicate that the vibration amplitude and angle of the vibration direction of the screen have a great influence on the velocity and

IOP Conf. Series: Materials Science and Engineering **843** (2020) 012024 doi:10.1088/1757-899X/843/1/012024

jump height of the materials. On the contrary, the vibration frequency and the inclined angle of the screening surface are insignificant.

For the machine working correctly with materials that are tough to screen, the authors suggest opting for the vibration frequency f = 13 Hz, the vibration amplitude A = 6.5 mm, the inclined angle of screening surface $\alpha_o = 6^0$, the angle of vibration direction of screen $\delta = 40^0$. The results of this study may be favorable for scientists in their field of vibrating screen design and calculations.

References

[1] Nguyen Van Xo and Golikov N S 219 Analysis of material particle motion and optimizing parameters of vibration of two-mass GZS vibratory feeder *Journal of Physics Conference Series* **1015(5)** 052020

[2] Hai Lin Dong Chu Sheng Liu Yue Min Zhao La La Zhao 2012 Review of the Development of Dry Coal Preparation Theory and Equipment *Advanced Materials Research* **619** 239-243

[3] Nguyen K L, Gabov V V, Doan V G and Pham V T 2019 Study of the influence of structure and parameters of loading and transporting devices of a cleaning combine on the efficiency of coal loading *Journal of Physics: Conference Series (JPCS)* **1384** 012036

[4] Nguyen Van Xo, Nguyen Khac Linh and Dinh Gia Nghiem 2019 Study the dynamics and analyze the effect of the position of the vibration motor to the oscillation of vibrating screen *Journal of Physics: Conference Series* **1384** 012035

[5] Chuan Guang Ding Fang Zhen Song Bo Song Xiu Hua Men 2011 Simulation Analysis of Dynamic Characteristics on Three-Body Vibrating Screen *Applied Mechanics and Materials* **105-107** 311-315

[6] Nguyen Van Xo 2018 Study kinetics, dynamics and parametric technology to improve the working efficiency of linear vibrating screen *Earth Sciences and Natural Resources for Sustainable Development* 44-48

[7] Hong Han Zehui Jiang Zehui Jiang Rui Zhang Jing Lyu 2016 The influence of air on period doubling motion in vertically vibrated grains beds *Physica A: Statistical Mechanics and its Applications* **403** 265–275

[8] Garcimartín A Maza D Ilquimiche JL Zuriguel I 2002 Convective motion in a vibrated granular layer *Phys Rev E Stat Nonlin Soft Matter Phys* **65** 031303

[9] Liu Bang Chun Liu Feng Liu Jie 1989 *Vibrating screen Vibrating feeder Design and adjustment of vibrating conveyor* (Chemical Press China)

[10] WangFeng WangHoa 1998 *Screening machinery* Mechanical Industry Press, Beijing China

[11] Wang Xin Wen 2003*Theoretical Analysis of Collision Motion between Particles of Material Group and Vibrating Bed Mining machine* (China)

[12] Xiao-mei He Chu-sheng Liu 2009 Dynamics and screening characteristics of a vibrating screen with variable elliptical trace *Mining Science and Technology (China)* **19** 508-513

[13] Xu Yong Sun Qi Cheng at al. 2003 *Research Progress of Particle Discrete Element Method. Progress in mechanics* (China)

[14] Zhu H P, Zhou Z Y, Yang R Y et al. 2007 *Discrete particle simulation of particulate systems: Theoretical developments*. Chemical Engineering Science **62(13)** 3378-3396

[15] Zhu H P , Zhou Z Y, Yang R Y et al. 2008 Discrete particle simulation of particulate systems: A review of major applications and findings *Chemical Engineering Science* **63** 5728-5770

[16] Wen Bang Chun, Liu Feng Qiao 1982 *Theory and Application of Vibration Machinery*(Beijing Machinery Industry Press)

AUTHOR INDEX